Earth and Environmental Sciences Library

Series Editors
Abdelazim Negm, Faculty of Engineering, Zagazig University, Zagazig, Egypt
Tatiana Chaplina, Antalya, Türkiye

Earth and Environmental Sciences Library (EESL) is a multidisciplinary book series focusing on innovative approaches and solid reviews to strengthen the role of the Earth and Environmental Sciences communities, while also providing sound guidance for stakeholders, decision-makers, policymakers, international organizations, and NGOs.

Topics of interest include oceanography, the marine environment, atmospheric sciences, hydrology and soil sciences, geophysics and geology, agriculture, environmental pollution, remote sensing, climate change, water resources, and natural resources management. In pursuit of these topics, the Earth Sciences and Environmental Sciences communities are invited to share their knowledge and expertise in the form of edited books, monographs, and conference proceedings.

Mustafa El-Rawy · Abdelazim Negm
Editors

Managed Aquifer Recharge in MENA Countries

Developments, Applications, Challenges, Strategies, and Sustainability

Springer

Editors
Mustafa El-Rawy
Department of Civil Engineering
Faculty of Engineering, Minia University
Minia, Egypt

Abdelazim Negm
Department of Water and Water Engineering
Faculty of Engineering, Zagazig University
Zagazig, Egypt

ISSN 2730-6674 ISSN 2730-6682 (electronic)
Earth and Environmental Sciences Library
ISBN 978-3-031-58763-4 ISBN 978-3-031-58764-1 (eBook)
https://doi.org/10.1007/978-3-031-58764-1

This Springer imprint is published by the registered company Springer Nature Switzerland AG
The registered company address is: Gewerbestrasse 11, 6330 Cham, Switzerland

Paper in this product is recyclable.

Preface

Water scarcity, exacerbated by limited natural water resources and a burgeoning population, presents a critical challenge in the Middle East and North Africa (MENA) region. The region's arid climate, coupled with increasing demands from urbanization and agriculture, underscores a pressing need for sustainable water management solutions. This book, "Managed Aquifer Recharge (MAR) in the MENA Countries," is written with the purpose of addressing these challenges by exploring MAR as a viable and innovative strategy to enhance water availability to face water security. Through methods like soil aquifer recharge, aquifer storage and recovery, rainfall harvesting, and riverbank filtration, MAR offers a means to replenish and sustain groundwater resources. The book aims to provide comprehensive insights into MAR techniques, their practical applications, and their potential to mitigate the impacts of water scarcity in the MENA region. By highlighting successful case studies and discussing the challenges and opportunities associated with MAR, this book serves as a crucial resource for policymakers, researchers, and practitioners in the field of water management, striving to balance the pressing water needs of the region's growing population with sustainable and efficient resource use.

This book has been written by 23 distinguished researchers, scientists, and water experts from diverse topic-related backgrounds and institutions. It stands as a testament to the collaborative expertise required to address the complex issue of water resource management via MAR. The contributors, each with extensive experience in water resources and specifically in groundwater management, ensure that the content is both authoritative and grounded in practical reality.

The book consists of 11 chapters, each exploring a different aspect of MAR and specific challenges present in the MENA region. It begins with an insightful examination of the challenges surrounding water resources in the region, providing a solid foundation for the topics that follow. The book then progresses to a series of chapters, each focused on a distinct MAR technique, including Soil Aquifer Recharge, Aquifer Storage and Recovery, River Bank Filtration, Rainfall Harvesting, and Infiltration Basins. These sections delve into both the technical details and the practical implementation of the MAR techniques within the MENA context, offering a comprehensive guide to their application.

An important feature of this book is its focus on case studies. These real-world examples offer a glimpse into the practical implementation of MAR strategies in the MENA region, reflecting the current challenges and the potential solutions that have been identified. Additionally, the book addresses MAR's environmental risks and impacts on groundwater quality, a crucial aspect of water resources sustainability. The concluding chapter brings together the most significant conclusions and recommendations derived from the preceding chapters but is not limited to them.

It is evident that the goal of the book is multifaceted. It provides an overview of the water resources challenges in the MENA region and the potential of MAR as a solution and highlights successful case studies and experiences that illuminate the path forward. It is designed to be a valuable resource for postgraduate students, academics, policymakers, and practitioners in water resources, offering theoretically robust insights grounded in real-world experiences.

Our collective effort in creating this book has been driven by the understanding that effective water resources management is crucial for the sustainability future of the MENA region. We strongly believe that the book will serve as a catalyst for further research, discussion, and action in the field of MAR and that it will contribute meaningfully and significantly to the ongoing efforts to address global water challenges.

The editors extend their profound appreciation to all authors and individuals who played a pivotal role in creating this unique and high-quality publication, a true beacon of knowledge and recent advancements in the field of MAR in the MENA region. Immense gratitude is owed to each author for their dedication and perseverance through the writing and revision processes, ensuring the manuscripts align with Springer's esteemed standards. Heartfelt thanks are also due to the series editor for his critical review and constructive comments in reviewing the chapters and the whole book again after the authors implemented all comments. Also, the editors want to thank Springer's team, whose comments, feedback, guidance, and meticulous follow-up and reviews have been invaluable.

The editors warmly welcome any feedback on this edition. Comments, critiques, suggestions for improvement, and ideas for new chapters in future editions, are all highly encouraged. For those wishing to contribute their insights, contact details for the editors are provided in several chapters throughout the book. Your input is crucial in enhancing the quality of future editions.

Minia, Egypt
Zagazig, Egypt
January 2024

Mustafa El-Rawy
Abdelazim Negm

Contents

An Overview of Water Resource Challenges and Managed Aquifer Recharge (MAR) in the MENA Region

Mustafa El-Rawy, Heba Fathi, Mohamed Wahba, and Abdelazim Negm

Abstract This chapter addresses the critical water scarcity issues in the Middle East and North Africa (MENA) region, which grapples with a dry climate and less than 1% of the world's freshwater, affecting its half-billion population. Factors exacerbating this scarcity include the region's low rainfall, below 200 mm annually, water quality issues, and excessive groundwater usage, exemplified by Saudi Arabia's annual extraction of 20 km^3. Additionally, rapid population growth and potential climate change impacts, possibly reducing rainfall by up to 20%, place further strain on the already limited water resources. With agriculture consuming from 75% to 90 of the water in MENA, the pressure intensifies. Current solutions such as desalination, as practiced in the UAE, and Egypt's wastewater reuse and sewage treatment, while effective, are energy-intensive and pose environmental challenges. This chapter proposes Managed Aquifer Recharge (MAR) as a sustainable solution to these challenges. MAR, recognized for its cost-effectiveness and environmental benefits, offers a promising approach for groundwater replenishment in arid regions. The chapter presents an overview of the dire water situation in the MENA region, the increasing demands due to population and economic growth, and explores the application of

M. El-Rawy (✉)
Civil Engineering Department, Faculty of Engineering, Minia University, Minia 61111, Egypt
e-mail: mustafa.elrawy@mu.edu.eg

Civil Engineering Department, College of Engineering, Shaqra University, Dawadmi 11911, Saudi Arabia

H. Fathi
College of Design and Architecture, Jazan University, Jazan 45142, Saudi Arabia
e-mail: hamansour@jazanu.edu.sa

M. Wahba
Civil Engineering Department, Faculty of Engineering, Mansoura University, Mansoura 35516, Egypt
e-mail: m_wahba@mans.edu.eg

A. Negm
Water and Water Structures Department Engineering, Faculty of Engineering, Zagazig University, Zagazig 44519, Egypt
e-mail: amnegm@zu.edu.eg

M. El-Rawy and A. Negm (eds.), *Managed Aquifer Recharge in MENA Countries*, Earth and Environmental Sciences Library,
https://doi.org/10.1007/978-3-031-58764-1_1

various MAR methods. It includes case studies demonstrating the effectiveness of MAR in sustainable water management in the MENA region. The chapter concludes by offering strategic recommendations for policymakers and water managers on integrating MAR into regional water resource planning and addressing water resource challenges in the MENA region.

Keywords Water resources · Water scarcity · Groundwater replenishment · Sustainable water management · MAR · MENA · Water quality

1 Introduction

The Middle East and North Africa (MENA) region is a diverse range of countries, including least developed countries like Sudan and Yemen, middle-income countries, and high-income Gulf oil-exporting countries [1]. The MENA region comprises 21 countries [2]. The MENA region is characterized by high temperatures for most of the year, these countries face high rates of evapotranspiration. In addition to relying on surface and groundwater, the region also utilizes unconventional sources such as desalination and the recycling of treated wastewater for its water needs [3].

The MENA region accounts for approximately 9.5% of the Earth's land surface (around 12.5 million square kilometers) [4]. This region is recognized as the most water-deprived area globally, with its per capita water resources averaging 550 m^3 annually. This figure is merely half of the 1,000 m^3/capita threshold that signifies water scarcity and slightly above the 500 m^3 benchmark for extreme water scarcity, as defined by the UN Water Stress Index [5]. Factors such as rapid population growth, urbanization, migration, and the expansion of irrigation and agricultural practices have escalated the region's water demand. Concurrently, the supply of water is dwindling, attributed to reduced rainfall and runoff and increased evapotranspiration, a consequence of climate change [6–8].

In 2018, the MENA region's population was reported at 487.7 million, as per the United Nations' 2017 data. With an annual growth rate of 1.7% recorded in 2017 [9], the region's population is projected to rise to 586 million by 2030 and 731 million by 2050, according to United Nations projections from 2017. The majority of countries within the MENA region are classified as low-income and middle-income for development assistance and financial lending.

Moreover, the MENA region, housing roughly 6% of the global population, possesses merely about 1% of the world's renewable freshwater resources, as identified in reports by the World Bank [9] and Kandeel [10]. This makes it the most water-deficient region on the planet. In 2011, the average per capita availability of Total Renewable Water Resources (TRWR) in the region was 819.8 m^3, but by 2016, this had dropped to around 600 m^3 [11]. For context, the worldwide average is approximately 6000 m^3. The current per capita freshwater availability in the MENA region is about 85% less than in the 1950s when it was around 4000 m^3/year [10].

The risks in the MENA region, which is already a very arid part of the world, are also increased by climate change. Furthermore, average temperatures have increased by 0.5 °C over the last century, and in certain regions of Sudan and North Africa, precipitation has dropped by up to 10% during the previous several decades. The MENA region receives 150–600 mm of precipitation annually, but its water resources are few, overused, contaminated, and politically sensitive [12, 13]. The Middle East receives between 3,000 and 5,000 h of sunlight annually, yet North Africa alone possesses enough renewable energy resources to provide all of Europe's energy demands [14].

Haddaoui and Mateo-Sagasta [3] declared that there would likely be less rainfall and that evapotranspiration rates would rise, increasing the need for agricultural water and reducing yields. For instance, under various climate change scenarios, the yields of Tunisia's rain-fed and irrigated grains will decrease by 5–11% until 2050. The water balance imbalance between the available water resources and the water needs has worsened in most MENA nations due to increased population growth, fast urbanization, and industry linked to rising living standards. Nevertheless, it is difficult to meet the sharp rises in water demand for household, industrial, and agricultural uses in the MENA region, particularly in areas with scarce water supplies [15]. To combat extended droughts, the MENA countries will need to discover practical solutions to climate change, such as water banking through aquifer storage and recovery during rainy years and implementing better-engineered water-harvesting techniques in larger regions not limited to the traditional places [7, 16]. Furthermore, machine learning techniques have been utilized to tackle the complexities associated with water resources management [17]. Additionally, these methods have also been applied to generate hazard maps for floods [18–21], and landslides [22].

This chapter will discuss water challenges and MAR applications only in specific MENA countries highlighted in the following sections, not all MENA countries.

2 Population Growth

The MENA region had a population of 487.7 million in 2018, growing at an average annual rate of 1.7% [9]. Projections estimate its population will increase to 586 million by 2030 and 731 million by 2050. Most MENA countries are categorized as low-income and middle-income for development and financial assistance. According to the World Bank [23], the MENA region's population is expected to have grown by 1.72%, from 0.3 billion in 2000 to 0.4 billion in 2017. It is projected that by 2030, there will be close to 0.5 billion residents. Figure 1 shows the population (thousands) in the year 2022 for each country in MENA. The population increase in the MENA region is faster than the 1.1% global rate. With almost 70% of the limited freshwater supply going toward agricultural output, these rates exacerbate the mismatch between availability and demand for water.

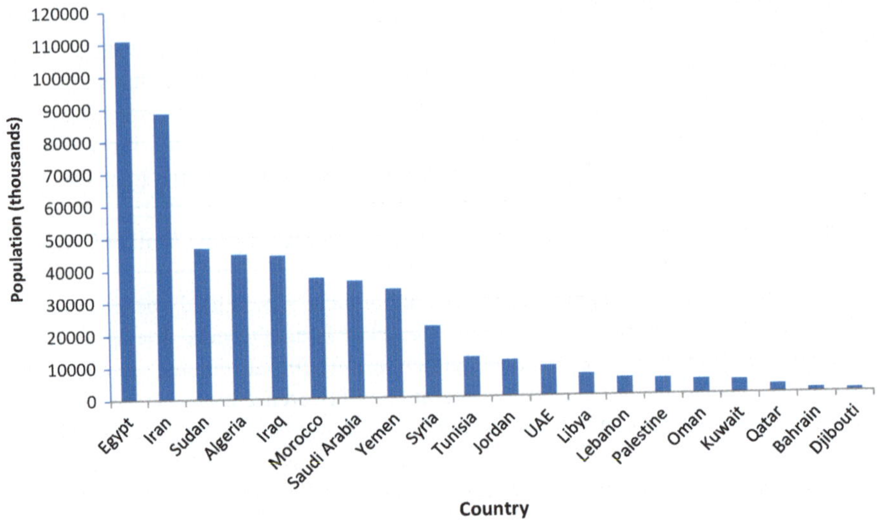

Fig. 1 Population (thousands) in the year 2022. *Source* The World Bank [23]

3 Groundwater Water Resources in the MENA Region

In the MENA region, groundwater renewability varies widely, from under 6 to 100% in some countries, influenced by factors like extraction rates, climate, hydrogeology, and water infrastructure. This variability highlights the critical concern of potential future water shortages, given water's fundamental role in human survival and development [24]. In 54% of countries in the MENA region, the primary water source is groundwater. The over-extraction of groundwater, a response to severe water stress, is expected to worsen over time [25].

Meanwhile, groundwater resources in the MENA region are a critical component of the area's overall water supply, although quantifying these resources can be complex due to varying methodologies and dynamic environmental factors.

The Arabian Peninsula, which includes Saudi Arabia, Kuwait, Bahrain, Qatar, the UAE, Oman, and Yemen, is predominantly reliant on non-renewable fossil aquifers. Saudi Arabia has an estimated 2.4 billion cubic meters (BCM) of renewable groundwater annually, but much of its water usage is from non-renewable sources [26]. Yemen, in the midst of a severe water crisis, has renewable groundwater resources estimated at 1.5 BCM per year [27], emphasizing the region's acute water scarcity.

The scenario varies significantly in the Levant, comprising Syria, Jordan, Lebanon, and Palestine. Jordan, for instance, is one of the world's most water-scarce countries, with renewable groundwater resources of about 0.5 BCM per year. Lebanon contrasts this with an estimated 0.8 BCM per year of renewable groundwater, reflecting its relatively richer water endowment [28, 29].

In addition, Iran and Iraq have their unique groundwater profiles. Iran's renewable groundwater resources are estimated at around 50 BCM per year, yet the country faces serious challenges due to over-extraction. Iraq, primarily dependent on the Tigris and Euphrates rivers, has less clearly defined groundwater figures but still considers it a vital resource [28, 29].

In Egypt, groundwater resources play a vital role in supplementing the primary water source, the Nile River. The Nubian Sandstone Aquifer System (NSAS), one of the largest fossil aquifers globally, extends beneath Egypt and holds an estimated 150,000 BCM of water, albeit with minimal recharge, making it largely non-renewable [30, 31]. Additionally, the Nile Valley and Delta's renewable groundwater resources, replenished by the Nile, irrigation return flow, and rainfall, contribute roughly 7.5 BCM annually [31, 32]. However, these figures are subject to fluctuation due to factors like Nile River flow, agricultural practices, and climatic conditions. Critical challenges such as seawater intrusion in the Nile Delta, which affects both the quantity and quality of groundwater, and potential impacts from the Grand Ethiopian Renaissance Dam on the Nile's flow and groundwater levels [33, 34], underscore the complexities in managing these vital resources in the face of growing demand and climate change. It's important to note that these figures are estimates and are subject to change due to various factors, including climate change, over-extraction, and policy changes. Additionally, the over-extraction and pollution of groundwater in many MENA countries are major concerns, potentially impacting these estimates significantly.

Figure 2a, b show a stark illustration of groundwater dependency across various countries in the MENA region, with the ratio of groundwater withdrawal to recharge revealing some concerning trends [28, 29]. In Saudi Arabia, for instance, the withdrawal rate is nearly four times the recharge rate, indicating a heavy reliance on groundwater that is not sustainable in the long term. Similarly, countries like the UAE and Libya exhibit extremely high withdrawal percentages compared to their recharge rates, at 689% and 587% respectively, which could lead to critical water shortages or the depletion of their groundwater resources. In contrast, Lebanon, and Morocco show much lower withdrawal rates compared to recharge, suggesting a more balanced use of groundwater. The withdrawal-to-demand ratio further highlights the dependency on groundwater to meet total water demand, with countries like Saudi Arabia, Oman, and Bahrain relying on groundwater for over 86%, 90%, and 92% of their total water demand in 2018, respectively. Conversely, Egypt stands out with a significantly lower dependency ratio, where groundwater accounts for just over 7% of the total water demand [28, 29]. These figures underscore the varying degrees of pressure on groundwater resources in the MENA region and the need for sustainable management practices to ensure water security.

There are two primary categories of aquifers in the MENA area:

1. Mega transboundary aquifers: These aquifers make up a sizeable amount of the Lower Sedimentary Column.
2. Surface aquifers: Over the MENA area, unconsolidated surface deposits range in age from the Pleistocene to the recent past, as well as in occurrence and

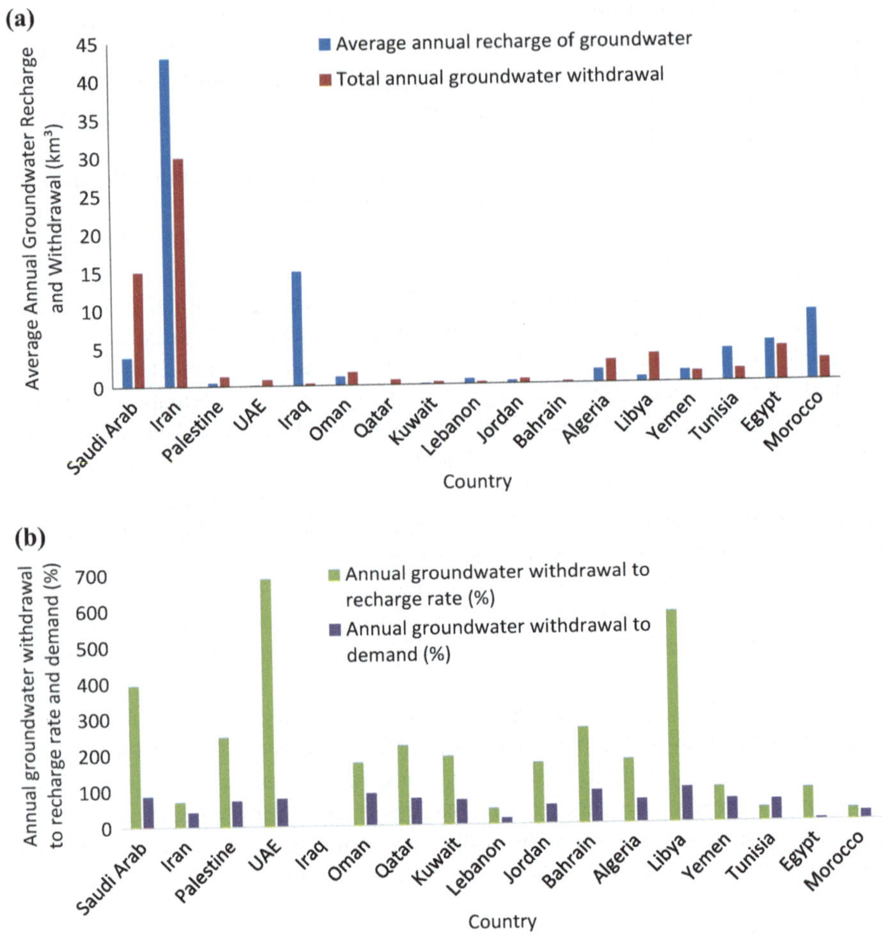

Fig. 2 Groundwater resources in countries of the MENA region: **a** average annual groundwater recharge and withdrawal (km^3); **b** Annual groundwater withdrawal to recharge rate and demand (%)

thickness. These lithological units can be broadly categorized into three groups: (a) eolian, (b) alluvial or fluvial, and (c) Sabkha deposits. These three units are primarily responsible for the distinctive landform characteristics that exist today. The hydrological regime of the region is significantly impacted by these entities [24].

The deep rock aquifer systems, both outcropping and near the surface, are currently unstable, experiencing a notably lower groundwater recharge rate than extraction. Groundwater abstraction from these aquifer systems is primarily utilized for agricultural purposes [35]. Despite being depleted or partially depleted, these aquifers possess the potential to store more water. This capacity is due to their extensive, unsaturated zones which were previously saturated with freshwater, along with their high transmissivity, substantial storage capacity, and the presence of fresh to moderately fresh local groundwater [36]. Due to the shared geological history within the MENA region, similar rock formations have given rise to aquifers crossing multiple countries' borders. This shared geology is the primary reason for the presence of transboundary aquifers such as the Paleogene aquifer, which stretches across the Arabian Peninsula, and the extensive Nubian Sandstone aquifer that lies beneath North African countries and Egypt. Similarly, the Basalt aquifer is a water source for both Jordan and Syria, and the Grand Erjs aquifer is a shared resource between Morocco, Tunisia, and Algeria. These shared aquifers underscore the need for cooperative water management between neighboring nations in the MENA region. Figure 3 shows the Map of groundwater resources in the MENA region, Source from [37], as reported by Sherif et al. [24].

In the MENA region, the concept of groundwater development stress, which measures groundwater abstraction as a percentage of annual recharge, is a critical metric in understanding the sustainability of water resources. The region, notorious for its arid climate and scarce water supplies, often sees groundwater abstraction rates that far exceed the natural recharge rates, leading to significant groundwater depletion. This imbalance is exacerbated by the region's reliance on groundwater for various purposes, including agriculture, which is the primary consumer of this resource. As a result, many aquifers in the MENA region are under severe stress (Fig. 4), with abstraction rates in some areas reaching alarming levels, far surpassing the annual recharge [38]. This overexploitation of groundwater resources poses a

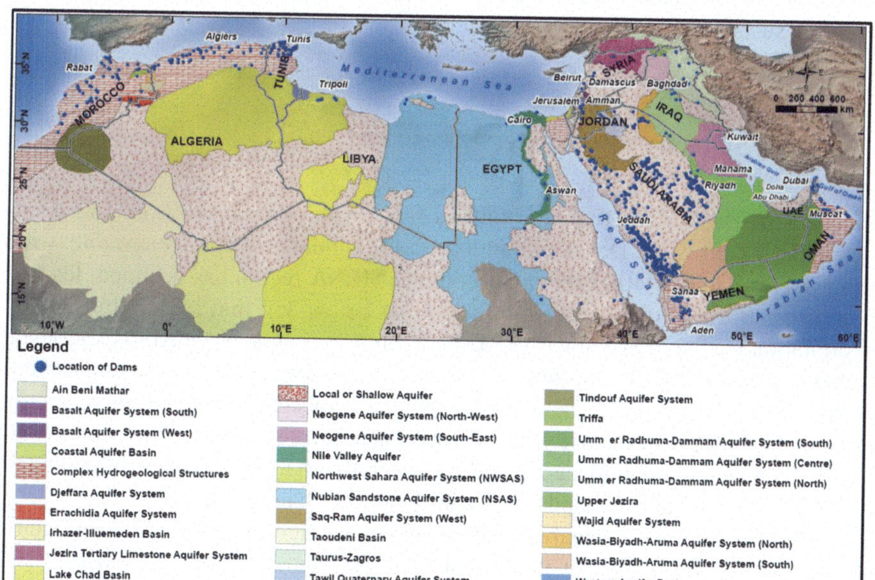

Fig. 3 Map of groundwater resources in the MENA region [24]

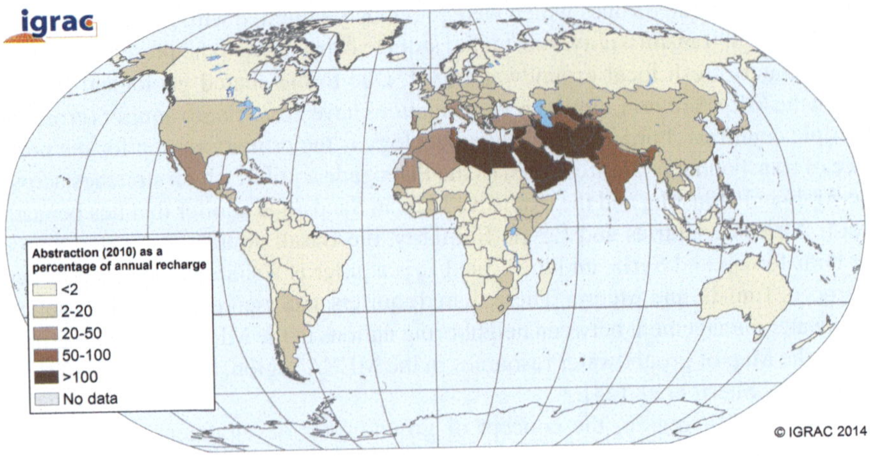

Fig. 4 Groundwater development stress map [38]

major challenge to water security in the region, threatening both the environment and the long-term viability of water supplies for future generations.

Nikraftar et al. [39] evaluated the sustainability of groundwater in the Middle East using data from GRACE/GRACE-FO. The findings indicate a mean decline in groundwater storage of 5.93 mm/year (37.29 km^3/year) in the region, with severe sustainability challenges affecting 59.14% of basins. The study also found that 34.41% of the basins are marginally unsustainable, while 6.45% demonstrate moderate sustainability.

4 Water Scarcity and Water Stress

Over half of the MENA region countries rank in the bottom 10% of 182 countries in the World Water Development Report for annual per capita renewable water resources availability [40]. Precipitation in the MENA region has declined, leading to drier conditions with average rainfall falling below 350 mm [41]. While global precipitation has increased since 1950, the MENA region is expected to experience further decreases [29, 41]. Jordan, for instance, might see a 30% reduction in rainfall by this century's end [42]. The MENA region is projected to become increasingly prone to droughts [43], intensifying existing water scarcity issues. This could result in significant economic impacts, including a potential 6 to 14% drop in annual GDP by 2050, reduced labor demand, and major land-use changes [9, 44].

The MENA region's renewable water resources average about 591,000 Mm^3 annually, accounting for 1.41% of the global total. Of this, 37%, or 222,000 Mm^3, is attributed to rivers flowing into the region from external areas [45]. Beyond renewable surface and groundwater, the region is also home to considerable non-renewable groundwater supplies. Additionally, MENA countries have varied entitlements to exploit unlimited quantities of seawater and brackish water.

Figure 5 presents a concerning trend of declining per capita annual renewable fresh water in cubic meters across MENA countries from 1970 to projections for 2025 [4]. Notably, countries like Iraq and Mauritania, which had relatively high-water resources per capita in 1970, have seen significant reductions over the decades. For instance, Iraq's per capita water availability has decreased from 8,478 m^3 in 1970 to a projected 2,338 m^3 by 2025. This trend is not isolated; almost every country listed shows a decrease, highlighting the increasing pressure on water resources in the region. Smaller countries with previously low water availability, such as Kuwait and the UAE, have seen further decreases to critical levels, indicating severe water scarcity challenges ahead. The overall MENA average has dropped from 1,752 m^3 per person in 1970 to an anticipated 551 m^3 by 2025, a reduction of more than two-thirds. This stark decline reflects the combined effects of population growth, agricultural expansion, industrialization, and potential impacts of climate change, including reduced rainfall and higher evaporation rates. Despite having higher per capita figures in 2000 compared to 1970, countries like Qatar and Saudi Arabia also face steep declines, emphasizing the widespread nature of this issue across the region. The data underscores the urgent need for sustainable water management practices, investment in water-saving technologies, and regional cooperation to address the looming water crisis in the MENA region [4]. In Yemen, the annual water demand for various uses, at 3.9 BCM, surpasses the combined renewable resources from surface water (1 BCM) and groundwater (1.5 BCM), totaling 2.5 BCM. The demand is expected to increase by the next decade to 4.5 BCM/year [27].

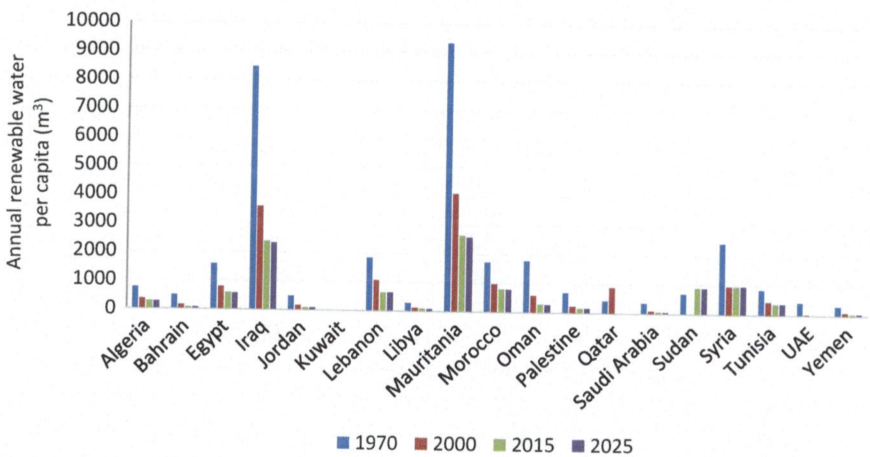

Fig. 5 Water resources available on a per-capita basis in countries of the MENA region [4]

5 Water Quality Issues

In the MENA (Middle East and Northern Africa) region, the quality of both surface water and groundwater is significantly impacted by various sources of pollution, exacerbated by natural conditions and human-induced factors [46]. Surface water sources such as rivers, lakes, and reservoirs are increasingly suffering from pollution. Industrial activities in the region contribute notably to this issue. Industries like petrochemicals, mining, and manufacturing discharge a range of pollutants, including heavy metals and toxic chemicals, directly into these water bodies [36]. Additionally, agricultural runoff, rich in fertilizers and pesticides, leads to the chemical contamination and eutrophication of surface waters [47].

Groundwater quality, which is crucial in this predominantly arid region, is also under threat. Over-extraction for agricultural, industrial, and domestic use has led to declining water levels and concentrations of contaminants [48]. Groundwater is further compromised by the seepage of pollutants from surface water, agricultural areas, and inadequately treated sewage, resulting in contamination with organic waste, pathogens, and other harmful substances [49]. On the other hand, water quality research necessitates a significant amount of data collection, laboratory analysis, data management, and testing [50].

Recycled wastewater is recognized as a non-conventional water resource in the MENA area due to a persistent shortage of water. However, a significant amount of the produced wastewater in this area is still discharged untreated into the sea or streams. In addition, the majority of the treated wastewater is released rather than recycled [35]. Also, the lack of adequate wastewater treatment infrastructure means that sewage is often discharged into the environment without sufficient treatment [3]. This not only contaminates water sources but also poses serious health risks and environmental concerns [51]. Countries like Saudi Arabia, Yemen, Egypt, and Jordan face particular challenges in this regard, with the combined effects of limited wastewater treatment, intensive agriculture, rapid industrialization, and excessive water extraction placing a considerable strain on water resources [13, 52].

To address these challenges, a comprehensive and multi-pronged approach is necessary. This should include improving agricultural practices to reduce runoff, managing and treating industrial discharges, upgrading wastewater treatment facilities, and implementing sustainable water extraction and management strategies [16, 53]. The efforts of researchers on this topic are notable, for example, Gabr et al. [54] suggested the construction of an entirely novel, sustainable, decentralized wastewater treatment system in Ar Riyadh, Saudi Arabia. This system may be an easy, affordable, decentralized, and energy-efficient method of treating urban wastewater sources, particularly for small-scale applications with up to 6000 users. The technique that was devised protects the environment and reuses treated water for gardening. The system is made up of a storage ground tank, a subsurface horizontal flow-built wetland, and a primary treatment (septic tank).

Also, methods that enable improved interpretation of water quality are becoming more and more necessary to guarantee efficient management and control of groundwater quality. Because of their statistical adaptability and simplicity of usage. One popular technique for keeping an eye on groundwater quality is to use water quality indicators (WQI) [55]. Consequently, Fathi and El-Rawy [56] evaluated Cairo's northeast groundwater quality and identified contamination hotspots associated with the two largest MENA wastewater treatment facilities, both having a daily capacity of more than three million m^3. The water quality index (WQI), which was initially calculated using multivariate statistical analysis, revealed a decline in water quality over the previous three decades. Furthermore, the use of multivariate statistical techniques like principal component analysis (PCA), cluster analysis (CA), and geographic information system (GIS) analysis has drawn more attention from researchers [57, 58]. The main factors impacting the regional distribution of groundwater quality could be identified through the combined use of GIS analysis and multivariate statistical techniques. For example, Gaagai et al. [59] used multivariate statistical analyses, backed by GIS techniques, to determine groundwater quality in the Doucen Plain, Algeria. The results demonstrated a useful and efficient method for evaluating groundwater quality and development. Eid et al. [60] also confirmed the same result in the Algerian Sahara's northeast, using multiple graphical and indexing approaches supported with GIS techniques. The state-of-the-art methodologies that have been published in the field of water quality and wastewater effluent evaluation are artificial neural networks (ANN), which have proven to be useful analysis tools to comprehend and predict [61]. This is observed in many studies in the region, El-Rawy et al. [62] created a machine learning framework to enhance effluent quality control of the El-Berka wastewater treatment plant, in Egypt which receives 600,000 m^3/day of wastewater, from approximately 3 million people. The plant's final effluent discharges into the Gabal El Asfar drain flows to the Bilbays drain, and then into the Bahr EI Baqar Drain and Lake Manzala; finally, it flows into the Mediterranean. Later, El-Rawy et al. [63] assessed how well six different artificial intelligence methods—including a machine-learning and integrated GIS method—performed in predicting the WQI in southern Saudi Arabia. The findings of the geographical distribution revealed that areas along the shore had greater levels of water pollution, which suggests that salinization and saltwater intrusion occur frequently there. Also, Ahmed et al. [64] used deep learning time series techniques (DLTS) and long short-term memory (LSTM) to assess and forecast groundwater quality for drinking purposes. WQI was carried out to evaluate key biological, chemical, and physical limitations of groundwater in Sohag, Egypt.

Consequently, strengthening regulatory frameworks and enhancing enforcement mechanisms are also crucial for protecting and improving the quality of water resources in the MENA region [15]. These steps are imperative not only for safeguarding environmental health but also for ensuring the socio-economic stability of the populations in these countries.

6 Climate Change Impacts

Climate change exerts a persistent and far-reaching impact on a substantial portion of the Earth's surface [65]. It significantly impacts water resources in the MENA region, both in terms of availability and quality, exacerbating the vulnerability of regional water security [6]. Rising temperatures, increased evapotranspiration, and decreases in precipitation and runoff due to climate change add further strain on these resources [7–9, 66]. Additionally, climate change has led to numerous devastating disasters, including widespread flooding [18, 19, 67, 68]. Climate change in the MENA is marked by regional climatic shifts and the response strategies of affected countries. The MENA region, home to 6% of the global population, produced 3.2 billion tons of CO_2, representing 8.7% of the world's greenhouse gas emissions in 2018, primarily from its energy sector [69, 70]. This region, rich in oil and natural gas, is highly vulnerable to climate impacts like droughts, heatwaves, and sea level rise.

The Intergovernmental Panel on Climate Change (IPCC) highlights the critical effects of climate change, such as varying temperatures and precipitation patterns, and increasing extreme weather events. MENA's arid conditions exacerbate these impacts, leading to concerns like low rainfall and high temperatures [71]. The IPCC predicts these conditions will worsen through the 21st century, potentially rendering parts of MENA uninhabitable by 2100 without significant GHG emission reductions [72].

The region's limited water and agricultural resources face further strain due to climate change, affecting national security and stability [73]. With over 60% of MENA's population living in high-water stress areas, countries are increasingly engaging in global climate initiatives such as the Paris Agreement and formulating national renewable energy policies.

Climate change economically impacts governments in the MENA region, as higher temperatures lower fiscal balance by 0.3% and raise debt by 1.87%, with rainfall changes having minimal effect. High greenhouse gas emissions could reduce fiscal balance by 7.3% and increase debt by 18% by the late 21st century, while lower emissions may lead to a fiscal reduction of 1.7–2.2 and a 5–6.3% debt increase [74].

Furthermore, a comprehensive investigation is currently underway by a collaborative team comprising members from NASA and Oxford University to delve into the influence of long-term climate variations on the early inhabitants and fauna of the Arabian Peninsula [75]. This study unveiled the presence of 10,000 lakes situated 100 m beneath the surface in the Empty Quarter of the Arabian Desert, indicative of an exceptionally moist climate that prevailed between 10,000 and 23,000 years ago. The findings suggest the potential revival of two major rivers in the Arabian Peninsula along with the possibility of the reemergence of two vast lakes functioning as water reservoirs. This inference was drawn from the dimensions of the Giant Lake, initially reaching a depth of 100 m (300 feet), and Ptolemy's Lake in Yemen, which is readily identifiable through satellite images on Google Earth [24].

7 Water Resource Challenges and Management

MENA area is the world's most water-stressed, with two-thirds of the countries still using more groundwater than their renewable freshwater supplies. Despite this, the area has the lowest water costs globally, substantial water subsidies (about 2% of Gross domestic product (GDP)), and a total water productivity that is just half of the global average [1]. MENA region has been reported to be among the most water-scarce regions. The amount of water extracted annually exceeds the renewable water available in the region [46].

However, Zekri and Al-Maamari [47] reported that irregular urban water delivery is still a problem in several Arab nations, including Yemen, Lebanon, and Sudan. Typically, residents store water in rooftop tanks to make adjustments. The system's high operating and maintenance expenses or a lack of water are the two primary causes of the supply interruption. In actuality, it takes a lot of energy and money to keep the pipes at a high pressure throughout the day. A benchmark of 1000 m^3 of renewable water resources per capita per year was set by the UN Food and Agriculture Organization (FAO), and 17 of the 21 MENA countries have surpassed it [51]. Iran, Iraq, Lebanon, and Morocco possess comparatively larger amounts of renewable water resources, with annual water levels exceeding 800 m^3 per person. Conversely, the Gulf Cooperation Countries (GCC), Yemen, Libya, Jordan, and Syria—have the least amount of renewable water resources accessible per person and are heavily dependent on desalination and groundwater depletion [21]. The MENA's problems with water resources will probably get worse in the future, mostly as a result of population growth and the effects of climate change.

Almost every country in the MENA region allocates over 70% of its renewable freshwater resources to agricultural activities, making it the predominant use of water. Following agriculture, the domestic sector ranks as the second-largest consumer of freshwater. In contrast, the industrial sector accounts for the smallest portion of freshwater usage, with less than 6% of the total consumption [12, 46].

The MENA region currently faces a water deficit, with its overall demand surpassing naturally accessible water by about 20%. This demand is projected to increase by 50% by 2050, as illustrated in Fig. 6, which outlines water supply sources and anticipated demand. The gap in water demand is likely to widen, mainly due to rising temperatures and growing populations [52]. Consequently, all countries in the MENA region will encounter considerable challenges in socioeconomic development due to water scarcity.

Groundwater is extensively used by all countries in the MENA region, a practice attributed to its easy access and the availability of energy resources for pumping, particularly in rural areas. Additionally, a common issue faced by most MENA countries is the low efficiency in water usage within irrigated agriculture. Compared to South Asia, for instance, this is lower, and many countries have turned to using their non-renewable resources for household, industrial, and agricultural reasons [15]. As a result, the MENA nations concentrated on finishing their development plans for sustainable water management, as was the case in Saudi Arabia, where

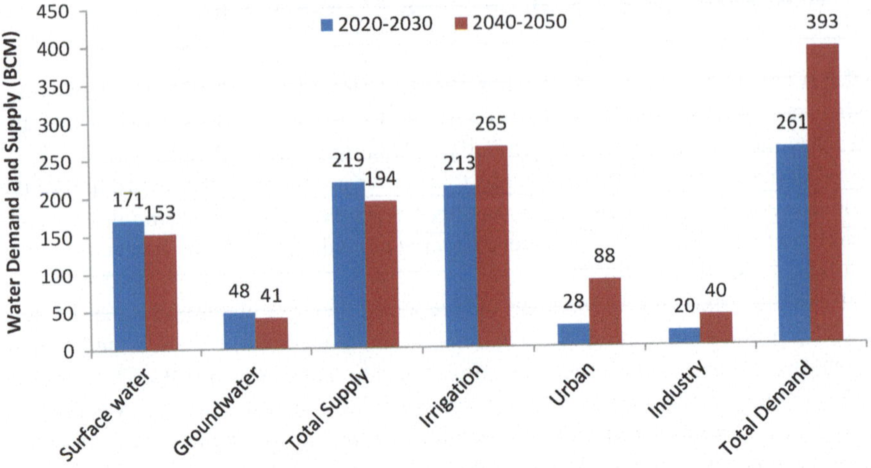

Fig. 6 Water demand and supply (BCM): **a** 2020–2030, **b** 2040–2050. Modified after [52]

90% of the water produced is used for irrigation. The national water strategy for Egypt intends to use 2400 million cubic meters (MCM) annually for agricultural purposes [1]. A better irrigation strategy (such as drip and sprinkler irrigation) and water-efficient plant seeds can minimize water waste and food loss while simultaneously increasing water availability, especially in light of the restricted resources [7]. Recharge ponds, soil aquifer treatment, recharge dams, aquifer storage transfer and recovery, and rooftop rainfall collection are some of the MENA region's aquifer recharge techniques [24, 76–82].

In addition, the MENA region heavily relies on desalinated water as its primary water source, with the GCC using desalination for almost 90% of its water. The United Arab Emirates operates the world's largest desalination facility for saltwater, with an annual capacity of about 10 MCM. In the region, 43% of wastewater is reused for irrigation [46]. However, of the 19 km^3 of municipal wastewater produced annually, only slightly more than 50% is treated. The remaining untreated wastewater, often due to non-functional or poorly maintained treatment facilities, is discharged into the environment. This leads to significant contamination of about 17 and 14% of the region's water bodies with pathogens and organic pollutants, respectively, posing risks to ecosystems and to those who use this water for irrigation or drinking [3].

Many countries in the MENA region have implemented regulations and standards for wastewater reuse, with others currently developing similar policies. For instance, Saudi Arabia's Ministry of Environment, Water, and Agriculture has set forth laws specifying the criteria for using recycled water, including quality and safety standards for treated wastewater in agriculture. Likewise, the Abu Dhabi Environment Agency in the United Arab Emirates has established regulations and guidelines for repurposing treated wastewater in various applications, such as industrial processes, landscaping, and irrigation [83]. The MENA region needs more comprehensive regulations for water reuse to provide greater assurance and confidence to investors and consumers. Adapting best practices to local conditions and learning from industrialized nations with established regulatory frameworks can be beneficial in this regard.

In Saudi Arabia, the management of treated wastewater is marked by the operation of 99 wastewater treatment plants (WWTPs), which collectively process an impressive annual volume of 1801.874 MCM. Daily, these facilities treat approximately 4.936 MCM of water, indicating a robust and consistent treatment infrastructure. Of the treated water, a daily volume of 852,307 m^3 is repurposed, reflecting a reuse rate of 17.26%. This percentage points to a significant, yet improvable, level of wastewater recycling, underscoring the potential for expanded reuse in the country's water resource management strategy. The current figures not only highlight the country's substantial investment in water treatment but also suggest avenues for enhancing water sustainability practices by increasing the proportion of recycled water. Further, the Ministry of Environment Water & Agriculture (SWPC) [84] has projected that Saudi Arabia's population will grow to 40.1 million by 2030, up from 33.6 million in 2018. This increase suggests that the total volume of wastewater will rise from 7.2 MCM per day in 2018 to 10.8 MCM per day by 2030 [85]. This anticipated growth underlines the importance of continued and enhanced efforts in managing and recycling wastewater to meet future demands.

Egypt, characterized by a semi-arid and dry climate [86, 87], necessitates alternative water resources to address deficiencies [77]. It, characterized as a semi-arid nation spanning a landmass of approximately one million square kilometers, occupies the southeastern region adjoining the Mediterranean Sea. The country grapples with water scarcity attributed to constrained natural water reservoirs, a burgeoning population, arid climatic conditions, amplified requisites across industrial, residential, and agricultural domains, as well as the necessity for potable water in reclaimed lands. The Nile River stands as the predominant source of water provision in Egypt, contributing approximately 55.5 BCM annually, representing over 97% of the total water supply [31, 32]. The remaining 3% is derived from renewable sources, fossil groundwater, and limited precipitation events [31, 88]. In numerous developing nations, the practice of utilizing untreated or diluted wastewater for irrigation has seen a notable rise owing to the limited availability of freshwater resources. The utilization of treated wastewater presents a reliable substitute resource for irrigation, crucial particularly in arid and semiarid regions where irrigation accounts for 80–85% of freshwater usage [89].

In addition, Oman faces severe water scarcity, significantly impeding various economic activities. To address this, the country is increasingly focusing on repurposing treated wastewater (TWW) to supplement its current water resources for various purposes. The Al-Khawd aquifer, situated in the Samail lower catchment (SLC), plays a crucial role in supplying water to Muscat, a city with around 1,210,480 inhabitants [90]. Currently, 68 public wells tap into this aquifer, drawing a minimum of about 6.83 MCM annually to support Muscat, alongside an additional 1.3 MCM used for agricultural practices. However, the reliance on desalinated water due to its high cost, energy demands, and environmental impacts necessitates a reevaluation of how the aquifer is managed [91]. Continued extensive usage of the aquifer to meet Muscat's water needs risks seawater intrusion and a decline in groundwater quality, aggravated by irregular rainfall patterns that sporadically replenish the aquifer. Managed Aquifer Recharge (MAR) utilizing treated wastewater is anticipated to enable the extraction of additional freshwater for urban use while minimizing harm to the aquifer [78].

The urban area of Muscat in Oman possesses an aquifer specifically designated for emergency situations. This aquifer is safeguarded and primarily utilized for urban needs, being refilled through subterranean movements and occasional sudden floods. Fluctuations in groundwater levels stem from various pressures, including periodic replenishment, extraction methods, and the discharge toward the ocean. Moreover, natural refilling remains unregulated, fluctuating, and unpredictable. When the hydraulic gradient intensifies toward the sea, discharge occurs, yet this can be reasonably managed by appropriately utilizing the aquifer [78.90]. Pumping activities can be regulated and adjusted to a certain extent [78].

From other perspectives, over 90% of Jordan's region experiences arid and semi-arid climatic conditions. The pressures exerted on the limited surface and groundwater reservoirs, primarily attributed to population expansion, agricultural advancements, and the influx of refugees, represent significant factors compelling planners within Jordan's Ministry of Water and Irrigation (MWI) to implement and expand MAR within their frameworks for water management and development. Among the existing agriculture practices is the utilization of treated wastewater (TWW) [81]. The Zarqa River in Jordan holds significant prominence as a primary contributor of surface water in Jordan, transporting over 78% of the total TWW [81]. Given the rising agricultural needs in the lower regions, the prospective repercussions on the Zarqa River will be pivotal concerning Jordan's water resources.

8 Managed Aquifer Recharge

The adoption of MAR has experienced substantial growth in dry and semi-dry regions due to its cost-effective, environmentally friendly, and robust approach to ensuring the sustainability and governance of water reserves [92].

The preparation for utilizing a dune field aquifer for MAR necessitates the estimation of hydraulic properties across an extensive geographic expanse [93]. Regarding water management, dune aquifers have been utilized for over six decades to store, purify, and retrieve water [94]. In regions with scant rainfall, sporadic rainfall events often result in flash floods that don't facilitate the replenishment of most aquifers, such as wadi aquifers [95]. Dune aquifers, however, have demonstrated efficacy in enabling natural recharge ([96]. There is a growing consideration for employing dune aquifers to retain captured stormwater or treated wastewater for future utilization [97]. Consequently, it is imperative to comprehensively comprehend the hydraulic characteristics of dune aquifers to evaluate their capacity for water storage and recovery using the principles of aquifer storage and recovery (ASR).

Moreover, the sustainability of groundwater in the MENA region amounts to less than half of the extraction rate, and this issue is exacerbated due to the influence of climate change [24]. Multiple aquifers span across borders; hence, overexploitation of these aquifers will have a widespread impact. Likewise, the encroachment of seawater into coastal aquifers has become a common occurrence in this area.

Meanwhile, water demand in MENA nations varies based on factors such as population size, standards of living, cultivated areas, crop types, irrigation methods, and industrial operations. Anticipated heightened occurrences of extreme precipitation events are expected to cause more frequent flash floods in the MENA region [98], the depletion of groundwater aquifers in is a potential focus for MAR, particularly in areas where infrastructure like recharge dams is available.

Additionally, the use of recharge dams for accumulating groundwater is a common practice in many countries globally. In the MENA region, most dams have been built primarily to enhance groundwater recharge. Even dams initially intended for protective purposes frequently function, at least partially, as groundwater recharge facilities. However, the effectiveness of these dams largely depends on the natural process of seepage and infiltration of water in the reservoir area, with limited application of MAR techniques [24].

The groundwater faces excessive exploitation due to a recharge rate from rainwater that significantly lags behind the rate of its extraction. In such circumstances, a viable strategy to improve the hydrodynamic and physicochemical aspects of groundwater involves the direct replenishment of water-table aquifers using rainwater collected behind dams [99].

Saudi Arabia has a long history of using dam construction for groundwater recharge and flood mitigation. The Hijaz region, particularly around Taif and Khaybar, is home to over twenty ancient dams, as documented [100]. Al-Rāshid [101] discussed the historic Wadi Al-Khanaq dam, located 15 km east of Madina, built between 661 and 679 AD. In more recent times, the 1956 construction of the Ikrimah dam in Taif, with a capacity of 0.5 MCM, marked the beginning of modern dam building in Saudi Arabia. Studies on groundwater recharge in the country [102, 103] indicate a significant increase in dam construction over the past 70 years. For instance, by 2006, the number of dams rose to 230, with heights ranging from 3 to 106 m [104]. By 2014, this number had doubled to 482 dams, distributed across various provinces and offering a total storage capacity of 2.08 billion cubic meters (BCM). Currently, Saudi Arabia has 563 dams with a combined storage capacity of 2.59 BCM, as shown in Fig. 7 [104].

The evolution of dam construction in Saudi Arabia, as highlighted by the steady increase in the number and capacity of existing dams, underscores the nation's commitment to managing water resources. However, despite these efforts, sustaining groundwater reservoirs remains a critical concern. MAR emerges as a proposed solution to address this challenge and enhance water resources sustainability. MAR involves deliberately directing surface water or treated wastewater into aquifers, essentially sustainable groundwater supplies. Implementing MAR strategies could

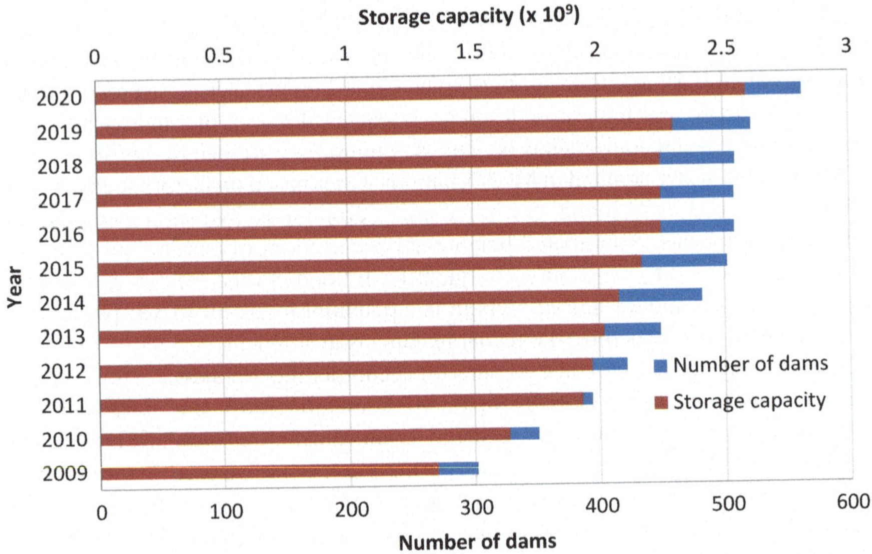

Fig. 7 The extant reservoirs within Saudi Arabia and their cumulative storage capability [104]

play a pivotal role in nurturing and fortifying the nation's groundwater reservoirs, thereby fostering long-term sustainability in Saudi Arabia's water resources management.

The Soil Aquifer Treatment (SAT) system represents a natural and advanced approach to treating partially treated wastewater without the use of chemical additives [77]. In arid and semi-arid regions, a viable and efficient tertiary treatment method involves the use of natural dilution, proving economically feasible for water reuse [105]. This technique entails the introduction of partially treated wastewater into soil layers, which subsequently enhances water quality [106]. The recharged water travels vertically through the unsaturated soil layers (vadose zone) and then horizontally within the aquifer towards the recovery wells. The vadose zone functions as a filtering mechanism, employing various biochemical and physical soil processes to eliminate a broad spectrum of contaminants [107]. SAT can address primary, secondary, or tertiary effluents and serves as a crucial strategy for conserving and effectively managing natural groundwater reservoirs [108]. Significant advantages of the SAT system include the potential for storing treated wastewater in the aquifer [109] and minimizing water evaporation from stored reserves [110].

In Saudi Arabia, the depletion of groundwater is occurring rapidly due to substantial withdrawals of water. Natural replenishment through rainwater infiltration is insufficient in this region. Consequently, partially treated wastewater is employed as an artificial means to recharge the groundwater. Enhancing the SAT system can significantly improve the process by enabling the infiltration of partially treated sewage into the soil, allowing it to percolate downward and recharge the groundwater to a greater extent [111].

9 Conclusions

The MENA region, characterized by its arid climate, faces significant challenges in water resource management. With less than 1% of the world's freshwater resources and about 6% of its population, the region is one of the most water-scarce in the world. In many areas, average annual precipitation is below 200 mm, contributing to the scarcity. Groundwater, which constitutes a major source of water in the region, is being depleted at an alarming rate, with countries like Saudi Arabia extracting up to 20 km^3 per year. High rates of population growth further exacerbate the reliance on non-renewable water sources; the region's population, currently around 562.71 million, and is expected to continue rising (https://www.statista.com/forecasts/1377703/mena-total-population). This increasing population, coupled with the effects of climate change, including an expected decrease in rainfall by up to 20% in some areas, adds immense pressure on the already limited water resources. Agricultural practices, consuming about 75% to 90 of water in the MENA region (https://arabdevelopmentportal.com/blog/water-consumption-agriculture), further strain these resources. Desalination has become a key strategy to meet water demands, with countries like the United Arab Emirates producing over 1.7 billion cubic meters of desalinated water annually. Egypt has expanded efforts in reusing agricultural wastewater, as well as in establishing sewage treatment plants. Additionally, there has been an expansion in the construction of desalination plants using seawater. However, these methods are energy-intensive and contribute to environmental concerns. Water scarcity in the MENA region limits water supply for domestic and agricultural use, posing a significant obstacle to the region's economic progress and stability. MAR provides a solution for replenishing and preserving groundwater resources through techniques like soil aquifer recharge, aquifer storage and recovery, rainwater harvesting, and riverbank filtration. In arid and semi-arid regions, MAR has grown considerably due to its cost-effective, environmentally sustainable, and dependable approach, aiding in the sustainable management of water reserves.

10 Recommendations

Several diverse strategies can be proposed to address water resource challenges and to implement Managed Aquifer Recharge (MAR) in the Middle East and North Africa (MENA) region. Below are a few of these recommendations:

- Develop and implement policies that promote sustainable water management practices. This includes regulations for water extraction, efficient water use, and allocation strategies that consider both social and environmental needs.
- Improve and invest in water infrastructure to reduce losses from leakages, enhance distribution systems, and expand MAR facilities. This includes the construction of dams, reservoirs, and rainwater harvesting systems.

- Encourage the adoption of water-saving technologies in agriculture, industry, and households. This might involve drip irrigation, efficient water fixtures, and practices that reduce water wastage.
- Conduct educational programs to raise awareness about the importance of water conservation. Training programs can also enhance local capacity in MAR techniques and management practices.
- Invest in research and development of new technologies for MAR, groundwater monitoring systems, and climate-resilient water management strategies tailored to the MENA region's specific challenges.
- Foster international and regional cooperation among countries in the MENA region for shared water management, including sharing knowledge, technologies, and resources.
- Involve local communities in decision-making processes related to water management to ensure sustainable and culturally acceptable solutions.
- Develop strategies that anticipate and adapt to climate change impacts on water resources. This might include adjusting MAR techniques to changing precipitation patterns and temperature variations.
- Establish robust monitoring and evaluation mechanisms to assess the effectiveness of MAR projects and overall water management strategies. Regular assessments can guide necessary adjustments and improvements.
- Combine traditional water management practices with modern MAR techniques to maximize efficiency and sustainability.

References

1. Marcel A, Enciso SR, Boubaker BB, Jonathan B, Emily C, Cicerone G, Katia C, De Cagno F, Koen D, Annelies D, Armelle E (2018) OECD-FAO Agricultural Outlook 2018–2027. In: OECD-FAO Agricultural Outlook 2018–2027, pp 1–112. https://doi.org/10.1787/agr_outlook-2018-en
2. Zekri S (ed) (2020) Water policies in MENA countries. Springer Cham. https://doi.org/10.1007/978-3-030-29274-4
3. Haddaoui I, Mateo-Sagasta J (2021) A review on occurrence of emerging pollutants in waters of the MENA region. Environ Sci Pollut Res 28:68090–68110. https://doi.org/10.1007/s11356-021-16558-8
4. FAO (Food and Agriculture Organization of the United Nations) (2022) Land area. Estimated average for the MENA region. AQUASTAT Database. https://www.fao.org/aquastat/statistics/query/index.html. Accessed 05 Oct 2023
5. Frascari D, Zanaroli G, Motaleb MA, Annen G, Belguith K, Borin S, Choukr-Allah R, Gibert C, Jaouani A, Kalogerakis N, Karajeh F (2018) Integrated technological and management solutions for wastewater treatment and efficient agricultural reuse in Egypt, Morocco, and Tunisia. Integr Environ Assess Manag 14(4):447–462. https://doi.org/10.1002/ieam.4045
6. IPCC (International Panel on Climate Change) (2021) Technical summary. In: Masson-Delmotte V, Zhai P, Pirani A, Connors SL, Péan C, Berger S, Caud N, Chen Y, Goldfarb L, Gomis MI, Huang M, Leitzell K, Lonnoy E, Matthews JBR, Maycock TK, Waterfield T, Yelekçi O, Yu R, Zhou B (eds) Climate change 2021: the physical science basis. Contribution of Working Group I to the Sixth Assessment Report of the Intergovernmental Panel on Climate

Change Cambridge. Cambridge University Press, United Kingdom, pp 33−144. https://doi.org/10.1017/9781009157896.002
7. El-Rawy M, Batelaan O, Al-Arifi N, Alotaibi A, Abdalla F, Gabr ME (2023) Climate change impacts on water resources in arid and semi-arid regions: a case study in Saudi Arabia. Water 15(3):606. https://doi.org/10.3390/w15030606
8. El-Rawy M, Fathi H, Zijl W, Alshehri F, Almadani S, Zaidi FK, Aldawsri M, Gabr ME (2023) Potential effects of climate change on agricultural water resources in Riyadh Region, Saudi Arabia. Sustainability 15(12):9513. https://doi.org/10.3390/su15129513
9. World Bank (2018) Beyond scarcity: Water security in the Middle East and North Africa. MENA Development Report. World Bank, Washington, DC. https://openknowledge.worldbank.org/handle/10986/27659
10. Kandeel A (2019) Freshwater resources in the MENA region: risks and opportunities. Middle East Institute. https://www.mei.edu/publications/freshwater-resources-mena-region-risks-and-opportunities
11. Sadik AK, El-solh M, Saab N (eds) (2014) Arab Environment 7: food security challenges and prospects: report of the Arab Forum for the Environment and Development. Arab Forum for the Environment and Development. https://wedocs.unep.org/20.500.11822/9621
12. Al-Sa'ed R, Sayadi S, Ghata A et al (2009) Advancing membrane technologies for wastewater treatment and reclamation in selected Arab MENA countries. Desalin Water Treat 4:287–293. https://doi.org/10.5004/dwt.2009.496
13. Zyadin A (2013) Water shortage in MENA region: an interdisciplinary overview and a suite of practical solutions. J Water Resour Protect 5:49–58. https://doi.org/10.4236/jwarp.2013.54A008
14. Abdelrassoul RA (1998) Potential for economic solar desalination in the Middle East. Renew Energy 14(1–4):345–349. https://doi.org/10.1016/S0960-1481(98)00088-3
15. Tropp H, Jagerskog A (2006) Water scarcity challenges in the Middle East and North Africa (MENA). Human Development Paper. UNDP. https://hdr.undp.org/content/human-development-report-2006
16. Hassen W, Mehri I, Beltifa A, Giorgia Potortì A, Khellaf N, Amer R, Van Loco J, Hassen A, Di Bella G, Khdary NH, Ben Mansour H (2022) Chemical and microbiological assessment of wastewater discharged along the Mediterranean Sea. Sustainability 14(5):2746. https://doi.org/10.3390/su14052746
17. Miro ME, Groves D, Tincher B, Syme J, Tanverakul S, Catt D (2021) Adaptive water management in the face of uncertainty: integrating machine learning, groundwater modeling and robust decision making. Clim Risk Manag 34:100383. https://doi.org/10.1016/j.crm.2021.100383
18. El-Rawy M, Elsadek WM, De Smedt F (2023) Flood hazard assessment and mitigation using a multi-criteria approach in the Sinai Peninsula, Egypt. Nat Hazards 115:215–236. https://doi.org/10.1007/s11069-022-05551-0
19. El-Rawy M, Elsadek WM, De Smedt F (2022) Flash flood susceptibility mapping in Sinai, Egypt using hydromorphic data, principal component analysis and logistic regression. Water 14(15):2434. https://doi.org/10.3390/w14152434
20. Wahba M, El-Rawy M, Al-Arifi N, Mansour MM (2023) A novel estimation of the composite hazard of landslides and flash floods utilizing an artificial intelligence approach. Water 15(23):4138. https://doi.org/10.3390/w15234138
21. Wahba M, Sharaan M, Elsadek WM, Kanae S, Hassan HS (2024) Building information modeling integrated with environmental flood hazard to assess the building vulnerability to flash floods. Stoch Environ Res Risk Assess (Accepted and under publishing)
22. Youssef AM, Pourghasemi HR. Landslide susceptibility mapping using machine learning algorithms and comparison of their performance at Abha Basin, Asir Region, Saudi Arabia. Geosci Front 12(2):639–655. https://doi.org/10.1016/j.gsf.2020.05.010
23. The World Bank (2023) The World Bank database. https://data.worldbank.org/indicator/SP.POP.TOTL?locations=ZQ. Accessed 10 Sept 2023

24. Sherif M, Sefelnasr A, Al Rashed M, Alshamsi D, Zaidi FK, Alghafli K, Baig F, Al-Turbak A, Alfaifi H, Loni OA et al (2023) A review of managed aquifer recharge potential in the Middle East and North Africa Region with examples from the Kingdom of Saudi Arabia and the United Arab Emirates. Water 15(4):742. https://doi.org/10.3390/w15040742
25. Khater AR (2001) Intensive groundwater use in the Middle East and North Africa, Intensive use of groundwater: challenges and opportunities. https://www.researchgate.net/publication/327817354_Intensive_groundwater_use_in_the_Middle_East_and_North_Africa
26. Chowdhury S, Al-Zahrani M (2015) Characterizing water resources and trends of sector wise water consumptions in Saudi Arabia. J King Saud Univ Eng Sci 27(1):68–82. https://doi.org/10.1016/j.jksues.2013.02.002
27. van den Berg H, van Meijeren, SJ, Gevaert A, Bazin J (2021) Water availability in Yemen. United Nations Development Programme (UNDP). https://www.undp.org/yemen/publications/water-availability-yemen
28. Das A (2020) Assessing water resources sustainability in Middle East and North Africa (Master's thesis, A. Das). https://urn.fi/URN:NBN:fi:oulu-202006182597
29. ESCWA (Economic and Social Commission for Western Asia) (2020). https://data.unescwa.org/#country-data. Accessed 20 Dec 2019
30. El-Rawy M, De Smedt F (2020) Estimation and mapping of the transmissivity of the Nubian Sandstone Aquifer in the Kharga Oasis, Egypt. Water 12(2):604. https://doi.org/10.3390/w12020604
31. El-Rawy M, Abdalla F, El Alfy M (2020) Water resources in Egypt. In: The geology of Egypt. Springer, Cham, pp 687–711. https://doi.org/10.1007/978-3-030-15265-9_18
32. Negm AM, El-Rawy M (eds) (2022) Sustainability of Groundwater in the Nile Valley, Egypt. Springer Nature, Cham. https://doi.org/10.1007/978-3-031-12676-5
33. El-Rawy M, Moghazy HE, Eltarabily MG (2021) Impacts of decreasing Nile flow on the Nile valley aquifer in El-Minia governorate, Egypt. Alex Eng J 60(2):2179–2192. https://doi.org/10.1016/j.aej.2020.12.037
34. El-Rawy M, Negm AM (2022) Sustainability of groundwater in the Nile River Valley: is it possible after the construction of GERD?. In: Negm AM, El-Rawy M (eds) Sustainability of groundwater in the Nile Valley, Egypt. Earth and Environmental Sciences Library. Springer, Cham. https://doi.org/10.1007/978-3-031-12676-5_13
35. Abu-Madi MOR (2004) Incentive systems for wastewater treatment and reuse in irrigated agriculture in the MENA region, evidence from Jordan and Tunisia. CRC Press. ISBN 9058097021 (Taylor & Francis Group). http://es.ircwash.org/
36. Ajjur SB, Baalousha HM (2021) A review on implementing managed aquifer recharge in the Middle East and North Africa region: methods, progress and challenges. Water Int 46:578–604. https://doi.org/10.1080/02508060.2021.1889192
37. Bundesanstalt für Geowissenschaften und Rohstoffe WHYMAP World-Wide Hydrogeological Mapping & Assessment Programme. http://www-naweb.iaea.org/napc/ih/documents/WAVE/WHYMAP-IAEA-May2010.pdf. Accessed 25 Feb 2021
38. IGRAC (International Groundwater Resources Assessment Centre) (2014) Groundwater Development Stress Map. https://www.un-igrac.org/resource/groundwater-development-stress-map. Accessed 15 Nov 2023
39. Nikraftar Z, Parizi E, Saber M, Hosseini SM, Ataie-Ashtiani B, Simmons CT (2023) Groundwater sustainability assessment in the Middle East using GRACE/GRACE-FO data. Hydrogeol J. https://doi.org/10.1007/s10040-023-02717-3
40. Tropp H (2007) Groundwater challenges in the MENA Region. UNDP Water Governance Facility at Stockholm International Water Institute (SIWI). https://www.iemed.org/publication/groundwater-challenges-in-the-mena-region/
41. ESCWA (Economic and Social Commission for Western Asia of the United Nations) (2019) Moving towards water security in the Arab Region. United Nations publication issued by ESCWA, Beirut, Lebanon

42. Wehrey F, Fawal N (2022) Cascading climate effects in the Middle East and North Africa: adapting through inclusive governance. CEIP: Carnegie Endowment for International Peace. https://policycommons.net/artifacts/2267979/cascading-climate-effects-in-the-middle-east-andnorth-africa/3027645/. Accessed 18 July 2023
43. Driouech F, ElRhaz K, Moufouma-Okia W, Khadija A, Saloua B (2020) Assessing future changes of climate extreme events in the CORDEX-MENA region using Regional Climate Model ALADIN-Climate. Earth Syst Environ 4:477–492. https://doi.org/10.1007/s41748-020-00169-3
44. Taheripour F, Tyner WE, Sajedinia E, Aguiar A, Chepeliev M, Corong E, de Lima CZ, Haqiqi I (2020) Water in the balance: the economic impacts of climate change and water scarcity in the Middle East. World Bank, Washington, DC
45. Miller JE (2003) Review of water resources and desalination technologies (No. SAND2003-0800). Sandia National Lab. (SNL-NM), Albuquerque, NM (United States); Sandia National Lab. (SNL-CA), Livermore, CA (United States). https://doi.org/10.2172/809106
46. Rateb A, Scanlon BR, Fakhreddine S (2022) How severe is water stress in the MENA region? Insights from GRACE and GRACE-FO Satellites and global hydrological modeling. In: Al Saud MM (eds) Applications of space techniques on the natural hazards in the MENA region. Springer, Cham. https://doi.org/10.1007/978-3-030-88874-9_4
47. Zekri S, Al-Maamari A (2020) An overview of the water sector in MENA region. In: Zekri S (eds) Water policies in MENA countries. Global issues in water policy, vol 23. Springer, Cham. https://doi.org/10.1007/978-3-030-29274-4_1
48. Ibrahim H, Yaseen ZM, Scholz M, Ali M, Gad M, Elsayed S, Khadr M, Hussein H, Ibrahim HH, Eid MH, Kovács A (2023) Evaluation and prediction of groundwater quality for irrigation using an integrated water quality indices, machine learning models and GIS approaches: a representative case study. Water 15(4):694. https://doi.org/10.3390/w15040694
49. El-Rawy M, Fathi H, Abdelrady A, Negm AM (2022) Environmental impacts of treated wastewater contaminates on groundwater quality in the Nile River Valley, Egypt. In: Negm AM, El-Rawy M (eds) Sustainability of groundwater in the Nile Valley, Egypt. Earth and environmental sciences library. Springer, Cham. https://doi.org/10.1007/978-3-031-12676-5_5
50. Rajmohan N, Masoud MHZ, BaM N (2021) Assessment of groundwater quality and associated health risk in the arid environment, Western Saudi Arabia. Environ Sci Pollut Res 28:9628–9646. https://doi.org/10.1007/s11356-020-11383-x
51. Asaad S, Suleiman A (2023) Socio-demographic factors and treated wastewater reuse in the MENA region: insights and implications. Desalination 565:116830. https://doi.org/10.1016/j.desal.2023.116830
52. Mualla W (2018) Water demand management is a must in MENA countries… but is it enough. J Geol Resour Eng 6:59–64. https://doi.org/10.17265/2328-2193/2018.02.002
53. Bekele E, Page D, Vanderzalm J, Kaksonen A, Gonzalez D (2018) Water recycling via aquifers for sustainable urban water quality management: current status, challenges and opportunities. Water 10(4):457. https://doi.org/10.3390/w10040457
54. Gabr ME, El-Rawy M, Al-Arifi N, Zijl W, Abdalla F (2023) A subsurface horizontal constructed wetland design approach for wastewater treatment: application in Ar Riyadh, Saudi Arabia. Sustainability 15(22):15927. https://doi.org/10.3390/su152215927
55. Masoud MH, Rajmohan N, Basahi JM, Niyazi BA (2022) Application of water quality indices and health risk models in the arid coastal aquifer, Southern Saudi Arabia. Environ Sci Pollut Res 29:70493–70507. https://doi.org/10.1007/s11356-022-20835-5
56. Fathi H, El-Rawy M (2018) GIS-based evaluation of water quality index for groundwater resources nearby wastewater treatment plants, Egypt. Pollut Res 37(1):105–116
57. Elsayed S, Hussein H, Moghanm FS, Khedher KM, Eid EM, Gad M (2020) Application of irrigation water quality indices and multivariate statistical techniques for surface water quality assessments in the Northern Nile Delta, Egypt. Water 12(12):3300. https://doi.org/10.3390/w12123300

58. Taşan M, Demir Y, Taşan S (2022) Groundwater quality assessment using principal component analysis and hierarchical cluster analysis in Alaçam, Turkey. Water Supply 22:3431–3447. https://doi.org/10.2166/ws.2021.390
59. Gaagai A, Aouissi HA, Bencedira S, Hinge G, Athamena A, Heddam S, Gad M, Elsherbiny O, Elsayed S, Eid MH, Ibrahim H (2023) Application of water quality indices, machine learning approaches, and GIS to identify groundwater quality for irrigation purposes: a case study of Sahara Aquifer, Doucen Plain, Algeria. Water 15(2):289. https://doi.org/10.3390/w15020289
60. Eid MH, Elbagory M, Tamma AA, Gad M, Elsayed S, Hussein H, Moghanm FS, Omara AE, Kovács A, Péter S (2023) Evaluation of groundwater quality for irrigation in deep aquifers using multiple graphical and indexing approaches supported with machine learning models and GIS techniques, Souf Valley, Algeria. Water 15(1):182. https://doi.org/10.3390/w15010182
61. Hamada M, Adel Zaqoot H, Abu Jreiban A (2018) Application of artificial neural networks for the prediction of Gaza wastewater treatment plant performance-Gaza strip. J Appl Res Water Wastewater 5:399–406. https://doi.org/10.22126/arww.2018.874
62. El-Rawy M, Abd-Ellah MK, Fathi H, Ahmed AKA (2021) Forecasting effluent and performance of wastewater treatment plant using different machine learning techniques. J Water Process Eng 44:102380. https://doi.org/10.1016/j.jwpe.2021.102380
63. El-Rawy M, Batelaan O, Alshehri F, Almadani S, Ahmed MS, Elbeltagi A (2023) An integrated GIS and machine-learning technique for groundwater quality assessment and prediction in Southern Saudi Arabia. Water 15(13):2448. https://doi.org/10.3390/w15132448
64. Ahmed AKA, El-Rawy M, Ibraheem AM, Al-Arifi N, Abd-Ellah MK (2023) Forecasting of groundwater quality by using deep learning time series techniques in an Arid Region. Sustainability 15(8):6529. https://doi.org/10.3390/su15086529
65. Wahba M, Sharaan M, Elsadek WM, Kanae S, Hassan HS (2023) Categorization of urban basin according to the runoff depth: case study of Katsushika Ward and Edogawa City Basin, Japan. In: Ujikawa K, Ishiwatari M, Hullebusch EV (eds) Environment and sustainable development. ACESD 2022. Environmental science and engineering. Springer, Singapore. https://doi.org/10.1007/978-981-99-4101-8_10
66. Al-Maktoumi A, Zekri S, El-Rawy M, Abdalla O, Al-Wardy M, Al-Rawas G, Charabi Y (2018) Assessment of the impact of climate change on coastal aquifers in Oman. Arab J Geosci 11:501. https://doi.org/10.1007/s12517-018-3858-y
67. Elsadek WM, Wahba M, Al-Arifi N, Kanae S, El-Rawy M (2024) Scrutinizing the performance of GIS-based analytical Hierarchical process approach and frequency ratio model in flood prediction—Case study of Kakegawa, Japan. Ain Shams Eng J 15(2):102453. https://doi.org/10.1016/j.asej.2023.102453
68. Wahba M, Hassan HS, Elsadek WM, Kanae S, Sharaan M (2022) Prediction of flood susceptibility using frequency ratio method: a case study of fifth district, Egypt. In: 14th international conference on hydroscience and engineering, Izmir-Turkey
69. The World Bank Group (2020) Middle East & North Africa Climate Roadmap (2021–2025), Driving transformational climate action and green recovery in MENA. https://thedocs.worldbank.org/en/doc/6f868d4a875db3ef23ef1dc747fcf2ca-0280012022/original/MENA-Roadmap-Final-01-20.pdf
70. Abbass RA, Kumar P, El-Gendy A (2018) An overview of monitoring and reduction strategies for health and climate change related emissions in the Middle East and North Africa region. Atmos Environ 175:33–43. https://doi.org/10.1016/j.atmosenv.2017.11.061
71. El-Fadel M, Bou-Zeid ER (2003) Climate change and water resources in the Middle East: vulnerability, socio-economic impacts, and adaptation. https://doi.org/10.4337/9781781950258.00015
72. Pal J, Eltahir E (2016) Future temperature in southwest Asia projected to exceed a threshold for human adaptability. Nat Clim Change 6:197–200. https://doi.org/10.1038/nclimate2833
73. Waha K, Krummenauer L, Adams S, Aich V, Baarsch F, Coumou D, Fader M, Hoff H, Jobbins G, Marcus R, Mengel M (2017) Climate change impacts in the Middle East and Northern Africa (MENA) region and their implications for vulnerable population groups. Reg Environ Change 17(6):1623–1638. https://doi.org/10.1007/s10113-017-1144-2

74. Giovanis E, Ozdamar O (2022) The impact of climate change on budget balances and debt in the Middle East and North Africa (MENA) region. Clim Change 172:34. https://doi.org/10.1007/s10584-022-03388-x
75. Oxford University and NAS (2023). https://www.ox.ac.uk/news/2012-04-30-ancient-network-rivers-and-lakes-found-arabian-desert. Accessed 10 Dec 2023
76. Sánchez A, Rylance G (2018) When the taps run dry: water stress and social unrest revisited. Revista UNISCI. https://www.unisci.es/wp-content/uploads/2018/05/UNISCIDP47-3SANCHEZ.pdf
77. Amin HM, Gad AA, El-Rawy M, Abdelghany UA, Sadeek RA (2021) Assessment of wastewater contaminant concentration through the vadose zone in a soil aquifer treatment system. Appl Ecol Environ Res 19(3):2385–403. https://doi.org/10.15666/aeer/1903_23852403
78. Al-Maktoumi A, El-Rawy M, Zekri S (2016) Management options for a multipurpose coastal aquifer in Oman. Arab J Geosci 9:636. https://doi.org/10.1007/s12517-016-2661-x
79. El-Rawy M, Al-Maktoumi A, Zekri S, Abdalla O, Al-Abri R (2019) Hydrological and economic feasibility of mitigating a stressed coastal aquifer using managed aquifer recharge: a case study of Jamma aquifer, Oman. J Arid Land 11:148–159. https://doi.org/10.1007/s40333-019-0093-7
80. Al-Maktoumi A, Zekri S, El-Rawy M, Abdalla O, Al-Abri R, Triki C, Bazargan-Lari MR (2020) Aquifer storage and recovery, and managed aquifer recharge of reclaimed water for management of coastal aquifers. Desalin Water Treat 176:67–77. https://doi.org/10.5004/dwt.2020.25499
81. El-Rawy M, Zlotnik VA, Al-Raggad M, Al-Maktoumi A, Kacimov A, Abdalla O (2016) Conjunctive use of groundwater and surface water resources with aquifer recharge by treated wastewater: evaluation of management scenarios in the Zarqa River Basin, Jordan. Environ Earth Sci 75:1–21. https://doi.org/10.1007/s12665-016-5946-1
82. Abdelrady A, Sharma S, Sefelnasr A, El-Rawy M, Kennedy M (2020) Analysis of the performance of bank filtration for water supply in arid climates: case study in Egypt. Water 12(6):1816. https://doi.org/10.3390/w12061816
83. Van Steenbergen F, El Haouari N (2011) The blind spot in water governance: conjunctive groundwater use in the MENA countries. Water policy and law in the Mediterranean: an evolving nexus. UNESCO, Paris. https://unesdoc.unesco.org/ark:/48223/pf0000193015
84. SWPC (Saudi Water Partnership Company) (2020) SWPC-7-Year-planning-Statement 2020–2026. https://idadesal.org/wp-content/uploads/2020/09/SWPC-7-Year-planning-Statement-2020--2026-Eng.pdf
85. El-Rawy M, Fathi H (2023) Groundwater pollution sources and its quality in the Kingdom of Saudi Arabia: state of the art. In: The handbook of environmental chemistry. Springer, Berlin, Heidelberg. https://doi.org/10.1007/698_2023_1050
86. Wahba M, Mahmoud H, Elsadek WM, Kanae S, Hassan HS (2022) Alleviation approach for flash flood risk reduction in urban dwellings: a case study of Fifth District, Egypt. Urban Climate 42:101130. https://doi.org/10.1016/j.uclim.2022.101130
87. El-Rawy M, Ismail E, Abdalla O (2019) Assessment of groundwater quality using GIS, hydrogeochemistry, and factor statistical analysis in Qena Governorate, Egypt. Desalin Water Treat 162:14–29. https://doi.org/10.5004/dwt.2019.24423
88. El-Rawy M, Makhloof AA, Hashem MD, Eltarabily MG (2021) Groundwater management of quaternary aquifer of the Nile Valley under different recharge and discharge scenarios: a case study Assiut governorate, Egypt. Ain Shams Eng J 12(3):2563–2574. https://doi.org/10.1016/j.asej.2021.02.023
89. Mahmoud MA, El-Bably AZ (2017) Crop water requirements and irrigation efficiencies in Egypt. In: Negm AM (eds) Conventional water resources and agriculture in Egypt. The handbook of environmental chemistry, vol 74. Springer, Cham. https://doi.org/10.1007/698_2017_42
90. NCSI (2015) Statistical yearbook. National Center for Statistics and Information. 34, Government of Oman. https://www.ncsi.gov.om/Elibrary/LibraryContentDoc/ben_Statistical_Year_Book2015_0f93543f-4dbf-4294-9e86-eacf6e805f04.pdf

91. Zekri S, Maktoumi A, Abdalla O, Akil J, Charabi Y (2014) Hydrogeological and economical simulation: water provision in emergency situation for Muscat. Water Policy 16:340–357. https://doi.org/10.2166/wp.2013.187
92. Dillon P, Stuyfzand P, Grischek T, Lluria M, Pyne RD, Jain RC, Bear J, Schwarz J, Wang W, Fernandez E, Stefan C (2019) Sixty years of global progress in managed aquifer recharge. Hydrogeol J 27:1–30. https://doi.org/10.1007/s10040-018-1841-z
93. Lopez OM, Jadoon KZ, Missimer TM (2015) Method of relating grain size distribution to hydraulic conductivity in dune sands to assist in assessing managed aquifer recharge projects: Wadi Khulays Dune field, Western Saudi Arabia. Water 7(11):6411–6426. https://doi.org/10.3390/w7116411
94. Roosma E, Stakelbeek A (1990) Deep-well infiltration in north-Holland dune area. Groundwater 1990(28):778–782. https://doi.org/10.1111/j.1745-6584.1990.tb01992.x
95. Missimer TM, Drewes JE, Amy G, Maliva RG, Keller S. Restoration of wadi aquifers by artificial recharge with treated waste water. Groundwater 50:514–527. https://doi.org/10.1111/j.1745-6584.2012.00941.x
96. Dincer T, Al-Muhgrin A, Zimmerman U (1974) Study of infiltration and recharge through sand dunes in arid zones with special reference to stable isotopes and thermonuclear tritium. J Hydrol 23:79–109. https://doi.org/10.1016/0022-1694(74)90025-0
97. Maliva R, Missimer T (2012) Arid lands water evaluation and management. Springer, Berlin, Heidelberg. https://doi.org/10.1007/978-3-642-29104-3
98. Almazroui M, Islam MN, Balkhai KS, Sen Z, Masood A (2017) Rainwater harvesting possibility under climate change: a basin-scale case study over western province of Saudi Arabia. Atmos Res 189:11–23. https://doi.org/10.1016/j.atmosres.2017.01.004
99. IAH-MAR (2018) International Association of Hydrogeologists Commission on Managing Aquifer Recharge Report of Activities in 2018; International Association of Hydrogeologists, 2018. https://iah.org. Accessed 02 Dec 2023
100. Kay S (1978) Some ancient Dams of the Hejaz. In: Proceedings of the seminar for Arabian studies, London, UK. https://www.jstor.org/stable/41223204
101. Al-Rāshid SBA (2008) An early Umayyad Dam near Medina, Saudi Arabia. In: Proceedings of the seminar for Arabian studies. Archaeopress, London, UK, pp 265–275. https://www.jstor.org/stable/41223953
102. Missimer TM, Maliva RG, Ghaffour N, Leiknes T, Amy GL (2014) Managed aquifer recharge (MAR) economics for wastewater reuse in low population Wadi Communities, Kingdom of Saudi Arabia. Water 6(8):2322–2338. https://doi.org/10.3390/w6082322
103. Abdulrazzak M, Sorman A, Alhames A (1989) Techniques of artificial recharge from an ephemeral Wadi channel under extreme arid conditions. In: Artificial recharge of ground water. ASCE, pp 602–611
104. MRMWR (2008) Oman Water Resources Atlas, The Ministry of Regional Municipalities and Water Resources (MRMWR): Muscat, Sultanate of Oman
105. Bouwer H (2002) Artificial recharge of groundwater: hydrogeology and engineering. Hydrogeol J 10:121–142. https://doi.org/10.1007/s10040-001-0182-4
106. Donn M, Reed D, Vanderzalm J, Page D (2020) Assessment of E. coli attenuation during infiltration of treated wastewater: a pathway to future managed aquifer recharge. Water 12(1):173. https://doi.org/10.3390/w12010173
107. Amy G, Drewes J (2007) Soil aquifer treatment (SAT) as a natural and sustainable wastewater reclamation/reuse technology: fate of wastewater effluent organic matter (EfOM) and trace organic compounds. Environ Monit Assess 129(1–3):19–26. https://doi.org/10.1007/s10661-006-9421-4
108. Xanke J, Goeppert N, Sawarieh A, Liesch T, Kinger J, Ali W, Hötzl H, Hadidi K, Goldscheider N (2014) Impact of managed aquifer recharge on the chemical and isotopic composition of a karst aquifer: Wala reservoir, Jordan. Hydrogeol J 23(5):1027–1040. https://doi.org/10.1007/s10040-015-1233-6
109. Bhuiyan C (2015) An approach towards site selection for water banking in unconfined aquifers through artificial recharge. J Hydrol 523:465–474. https://doi.org/10.1016/j.jhydrol.2015.01.052

110. Ayuso-Gabella N, Page D, Masciopinto C, Aharoni A, Salgot M, Wintgens T (2011) Quantifying the effect of managed aquifer recharge on the microbiological human health risks of irrigating crops with recycled water. Agric Water Manag 99(1):93–102. https://doi.org/10.1016/j.agwat.2011.07.014
111. Elsheik MA, Elhamidy WK (2014) Wastewater reuse through soil aquifer treatment. J Eng Technol Res 2(1):25–35. https://www.scientiaresearchlibrary.com/archive/JETR-2014-2-1-018-25-35.pdf

Advancing Groundwater Management and Seawater Intrusion Monitoring in the Middle East and North Africa: Exploring the Potential of Nuclear Tracer Techniques

Mohamed A. S. Abdel Monem, Ismail A. El Ghandour, and M. F. Abdel-Sabour

Abstract Water scarcity in the Middle East and North Africa (MENA) region poses significant challenges to the sustainable management of groundwater resources. Factors such as population growth, non-rational water use, and the impact of climate change further exacerbate the situation. This chapter proposes the adoption of effective governance strategies for the limited groundwater resources to address these pressing issues. Groundwater management and monitoring of seawater intrusion pose significant challenges in the region due to increasing water demand, the threat of salinization, and sea level rise due to climate change. Stable and radioactive isotopes present a valuable and reliable tool for studying, understanding, and monitoring groundwater systems in the region. In arid and semi-arid climates, isotope techniques offer a practical and dependable methodology for detecting and quantifying groundwater recharge, as well as determining groundwater age and assessing pollution in both shallow and deeper aquifers. The use of environmental isotopes, including deuterium (^{2}H), tritium (^{3}H), oxygen-18 (^{18}O), and carbon-13 (^{13}C) and carbon-14 (^{14}C), provides insights into water provenance and assists in comprehending the age of groundwater. Tritium, with its high level of accuracy, is employed to trace groundwater history up to 50 years, while carbon-14 is utilized to investigate history spanning up to 40,000 years. Accurate assessment of hydrogeological parameters and aquifer characteristics is crucial for understanding groundwater origins and budgeting. Additionally, various environmental isotopes have been successfully employed to identify seawater intrusion in coastal aquifers and gain a comprehensive understanding of the processes influencing salt concentration in groundwater in seaboard areas. This chapter showcases the application of stable and radioactive isotopes in groundwater management within the Gulf Cooperation Council (GCC)

M. A. S. Abdel Monem · I. A. E. Ghandour (✉) · M. F. Abdel-Sabour
Soil and Water Department, Atomic Energy Authority, Nasr City, Cairo, Egypt
e-mail: ismailelghandour@yahoo.co.uk

M. A. S. Abdel Monem
Regional Office for the Near East and North Africa, FAO, Cairo, Egypt

M. El-Rawy and A. Negm (eds.), *Managed Aquifer Recharge in MENA Countries*, Earth and Environmental Sciences Library,
https://doi.org/10.1007/978-3-031-58764-1_2

countries, Mashreq countries, and countries in North Africa and Egypt. The findings and methodologies presented directly impact the livelihoods of diverse groundwater users in these regions. By incorporating isotope techniques into groundwater management strategies, MENA countries can make informed decisions regarding the utilization and protection of their valuable groundwater resources.

Keywords Environmental isotopes · Groundwater management · Sea water intrusion · MENA region

Abbreviations

ACSAD	Arab Center for Studies of Arid Zones and Dry Lands—ACSAD
DIC	Dissolved Inorganic Carbon
GCC	Gulf Cooperation Country
GDP	Growth Domestic Product
GIS	Geographic Information System
GISP	Greenland Ice Sheet Precipitation
GMWL	Global Meteoric Water Line
HTO	Hydrogen Tritium Oxide
IAEA	International Atomic Energy Authority
MENA	Middle East and North Africa
pMc	Percent of Modern Carbon
SGD	Submarine Ground water Discharge
SLAP	Standard Light Antarctic Precipitation
SMI	Seawater Mixing Index
SMOW	Standard Mean Oceanic Water
TU	Tritium Units
TDS	Total Dissolved Solid
UNESCO	United Nations Educational, Scientific and Cultural Organization
WISER	Water Isotope System for Data Analysis, conception, and Electronic Recovery

1 Introduction

The Middle East and North Africa (MENA) region heavily relies on groundwater to supplement scarce surface water for household and agricultural needs. Groundwater accounts for approximately 14% (45 billion cubic meters) of the total renewable water resources in the region, which is estimated at about 321 billion cubic meters [1]. In several countries , including Bahrain, Jordan, Lebanon, Oman, Tunisia, the

United Arab Emirates, and Yemen, groundwater contributes to more than 50% of total water withdrawals. In the Arabian Peninsula, it accounts for a significant 84% [1].

Non-renewable groundwater sources, also known as fossil groundwater, are found in large areas and at greater depths in the Arab region, particularly in the Sahara and the Arabian Peninsula, where over 65% of groundwater is non-renewable [2]. The accessibility and widespread presence of groundwater make it a heavily exploited resource in most MENA countries to meet the growing demand for water. This over-exploitation has led to the fast depletion of aquifer reserves, salinization, and deterioration of water quality, including seawater intrusion in coastal aquifers [2]. Figure 1 illustrates the extent of groundwater depletion in selected Arab countries, showing that the overexploitation of groundwater has resulted in a depletion of national wealth equivalent to approximately 2% of the GDP [3]. For example, in Tunisia, the shallow aquifers are being overused by 120%, resulting in subsidence of the water table in numerous areas and seawater intrusion in coastal aquifers in the northern part of the country [4]. In the Egyptian delta, groundwater extraction volume reaches approximately 5.6 billion cubic meters/year from 24,560 wells [5]. In Yemen, groundwater irrigation has increased by 1135% between 1970 and 2010 [6]. While the extraction of groundwater for economic activities may initially boost the GDP, the overuse of this resource poses a long-term threat, leading to negative consequences such as desertification and adversely impacting the country's natural capital, wealth, and the livelihoods of its people.

In the Arab countries, groundwater overutilization has dramatically increased over the past decades, primarily driven by the expansion of irrigated agriculture. As a result of excessive pumping, many groundwater resources are at risk of depletion. Climate change further compounds the issue, as it not only impacts the quantity of freshwater but also poses a threat to groundwater quality. The intrusion of seawater

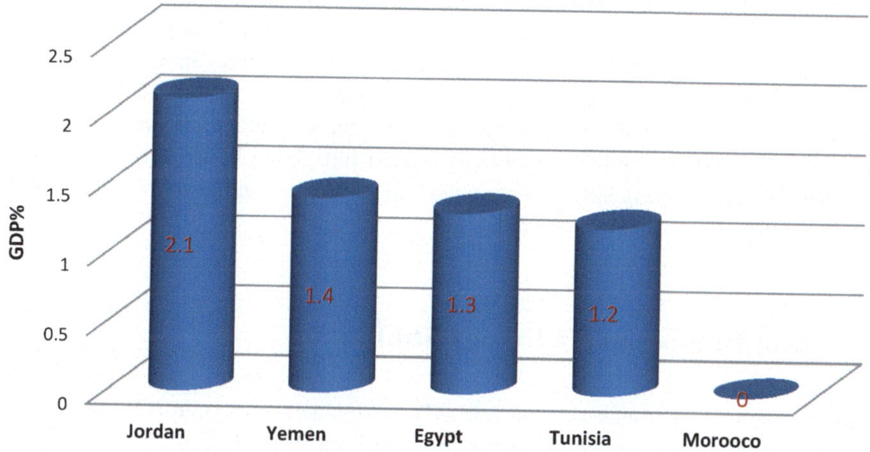

Fig. 1 Value of groundwater depletion in selected Arab countries. Modified after [3]

into coastal aquifers can contaminate freshwater supplies, affecting the availability of safe drinking water for millions of people in the Arab countries. Additionally, significant water volumes stored in deep aquifers in Libya, Tunisia, Egypt, Algeria, and the Arabian Peninsula are non-renewable resources, making their utilization unsustainable [7].

The Intergovernmental Panel on Climate Change [7] that sea levels could raise by 20–88 mm per year over the next century. This increase in sea levels, attributed to climatic changes, puts coastal areas at risk of flooding, with projections forecasting a rise of 0.5–1 m by the end of the 21st century. Studies suggest that more than 40% of the Nile Delta in Egypt could be vulnerable to flooding by 2050 [8]. If sea levels rise by 0.5 m, around 1800 km^2 of agricultural land and 12,000 km^2 of urban areas could be lost [9]. The predicted impacts of climate change on the Nile Delta suggest that sea level rise could cause primary or secondary effects, posing significant threats to soil fertility and groundwater quality by allowing seawater to infiltrate coastal aquifers, resulting in soil and water salinization. Groundwater salinization is a real threat to the livelihoods of millions of small-scale farmers in Egypt. It has been estimated that approximately 2.4 million feddan (about one million hectare) from irrigated agricultural areas in Egypt suffer from problems of waterlogging, salinization, and sodicity.

Managed Aquifer Recharge (MAR) has been recognized as an effective method for the sustainable management of groundwater in MENA countries. It has been extensively studied and implemented in various areas across the region, playing a crucial role in ensuring the long-term viability of groundwater resources [10–16]. However, the existing monitoring systems for groundwater resources, both at the national and shared levels, are inadequate and ineffective. There is a pressing need to prioritize data acquisition and development by adopting new technologies, equipment, and staff training. This will contribute to a better understanding of aquifer systems, which forms the basis for sound management practices [17–20]. The use of nuclear techniques provides an efficient, accurate, and reliable monitoring system for groundwater, enabling informed governance decisions and judgments. Recent applications of isotope hydrology cover several important areas, including the estimation of aquifer replenishment and storage, understanding the impacts of long-term groundwater overuse, managing shared groundwater aquifers, assessing freshwater discharge in Arabian countries, determining and mitigating hydrocarbon pollution in groundwater, studying soil moisture and salt movement in the unsaturated zone, and paleoclimate reconstruction [21].

2 Use of Isotopes in Hydrology Studies

Isotopes such as hydrogen (H), carbon (C), nitrogen (N), oxygen (O), and sulfur (S) are essential environmental tracers in hydrological, geological, and biological systems. These isotopes occur naturally in the environment and are used to trace groundwater origin, recharge processes, subsurface reactions, and reaction rates [22].

Environmental tracers have become a common tool in understanding groundwater flow processes, water budgets, recharge sources, chemical reactions, and deep aquifer compartments, as well as estimating groundwater age and validating flow and transport models [23]. Environmental tracers include radon, chlorofluorocarbons (CFCs), sulfur hexafluoride (SF6), noble gases, stable isotopes, and environmental tritium. Human activities, such as atmospheric nuclear weapons testing in the 1950s and 1960s, have also contributed to the presence of isotopes in the environment. These isotopes were released into the upper atmosphere and subsequently deposited on Earth through rainfall. Unintentional releases of isotopes, such as the Chernobyl accident in 1986, have also had an impact on the environment [22].

2.1 *Isotopes Classification*

As shown in Fig. 2 the isotopes used in groundwater studies could be classified as follows [24].

2.1.1 Stable Isotopes

Stable isotopes, including ^{2}H, ^{18}O, and ^{13}C, play a critical role in hydrological studies as tracers for water movement and cycling. A water molecule is made up of two hydrogen atoms and one oxygen atom, leading to 18 possible combinations such as

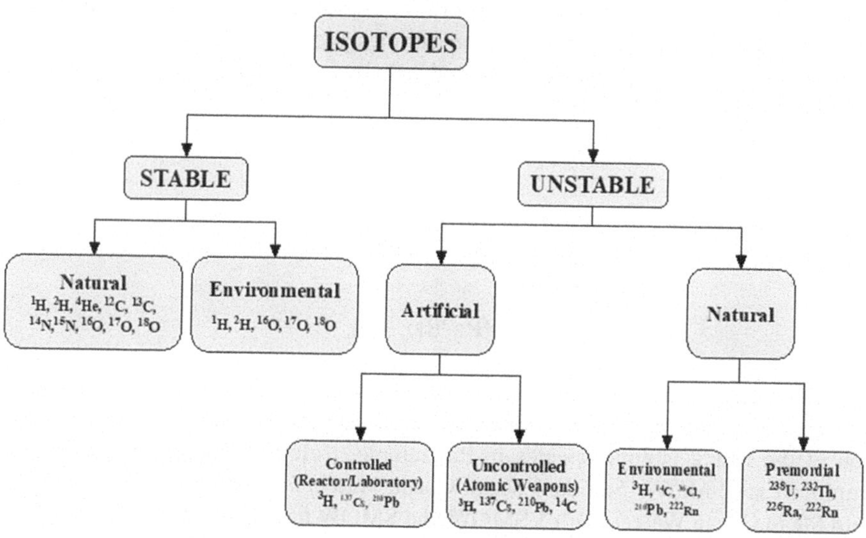

Fig. 2 Classification of isotopes [21]

$^1H^1H^{16}O$, $^1H^1H^{16}O$, $^1HD^{16}O$, $HD^{18}O$, $^1H^1H^{17}O$, and $^1HD^{17}O$. Each of these molecular configurations holds significant importance for understanding various hydrological processes. The distribution of these water molecules in nature adheres to a distinct pattern:

$$H_2^{16}O \sim 999680 \text{ ppm } (99.9680\%)$$

$$HD^{16}O \sim 320 \text{ ppm } (0.032\%)$$

$$H_2^{18}O \sim 2040 \text{ ppm}(0.204\%)$$

The measurement of stable isotopes is conducted by comparing their abundance ratios, essentially comparing the atomic mass of the heavier isotope to that of the lighter isotope. For instance, heavy water $2H_2{}^{16}O$ ($2H_2{}^{16}O$) has a mass of 20, whereas regular water $^1H_2{}^{16}O$ has a mass of 18. Likewise, the mass of a more substantial stable water molecule, $^2H_2{}^{18}O$, is 22. This discrepancy arises from the difference in neutron count. In natural waters and other materials, the absolute abundance of isotopes is not commonly measured. Rather, the focus is on measuring the relative difference in the ratio of the heavy to the light isotopes present in a sample, benchmarked against a reference standard. This relative difference is indicated by the Greek letter δ and is defined as follows:

$$\delta = \left(R_{sample} - R_{reference}\right)/R_{reference}$$

In this context, "R" denotes the ratios of the isotopes $^{18}O/^{16}O$ and D/H in water. Given that the discrepancies between the samples and their reference standards are often minimal, the δ values are presented in terms of per mill (‰), or parts per thousand (Table 1).

$$\delta(^0/_{00}) = \delta \times 1000$$

$$\delta(^0/_{00}) = [(R_s - R_r)/Rr] \times 10^3$$

$$\delta(^0/_{00}) = [(Rs/Rr) - 1] \times 10^3$$

A positive δ value signifies that the sample is enriched with the heavy isotope species in comparison to the reference, whereas a negative δ value indicates a depletion of the heavy isotope species in the sample. Reference standards commonly employed are SMOW (Standard Mean Ocean Water) and VSMOW (Vienna Standard Mean Ocean Water), with VSMOW and SMOW having identical ^{18}O contents, although the D content of VSMOW is 0.2‰ lower. Additionally, for calibrating

Table 1 Stable isotopes utilized in the study of groundwater environments [22]

Isotope	Ratio	Natural abundance (atoms/atoms)	Isotope	Ratio	Natural abundance (atoms/atoms)
^{2}H	$^{2}H/^{3}H$	1.55×10^{-4}	^{18}O	$^{18}O/^{16}O$	2.04×10^{-3}
^{3}He	$^{3}He/^{4}He$	1.38×10^{-6}	^{34}S	$^{34}S/^{32}S$	4.22×10^{-2}
^{6}Li	$^{6}Li/^{7}Li$	7.5×10^{-2}	^{37}Cl	$^{37}Cl/^{35}Cl$	0.242
^{11}B	$^{11}B/^{10}B$	0.80	^{81}Br	$^{81}Br/^{79}Br$	0493
^{13}C	$^{13}C/^{12}C$	$1,11 \times 10^{-2}$	^{87}Sr	$^{87}Sr/^{86}Sr$	0.7099
^{15}N	$^{15}N/^{14}N$	3.66×10^{-3}	–	–	–

oxygen and hydrogen isotopes, other reference standards such as SLAP (Standard Light Antarctic Precipitation) and GISP (Greenland Ice Sheet Precipitation) are used [22].

2.1.2 Radioisotopes

Radioisotopes are important tools for studying and monitoring groundwater. The most commonly used radioisotopes include Tritium (^{3}H) and Carbon 14 (^{14}C), among others, as shown in Table 2 [22]. The radioisotope of hydrogen, tritium (^{3}H), in the form of water molecule ($^{3}H_2O$), designated by the symbol ^{3}H or T, is still widely utilized for numerous hydrological studies.

Tritium naturally forms in the upper atmosphere through the interaction of cosmic rays with nitrogen molecules present in the air. It is also generated during nuclear weapons explosions and is a byproduct of nuclear reactors. Tritium and carbon-14

Table 2 Radioisotopes of cosmogenic and anthropogenic origin utilized in hydrological studies

Radioactive isotope	Half-life (a)	Atmospheric production		In situ production	
		Reaction	Initial volume of groundwater in liters	Reaction	Equilibrium content in 1L of groundwater
^{85}Kr	10.76	Anthropogenic	2.5×10^{-6} Bq	None	0
^{3}H	12.43	$^{14}N(n, t)$	1 Bq	$^{6}Li(n \cdot \alpha)$	2×10^{-5} q
^{39}Ar	269	$^{40}Ar\ (n \cdot 2n)$	8.5×10^{-3} Bq	$^{39}K(n \cdot P)$	$0.2 - 2 \times 10^{3}$Bq
^{14}C	5730	$^{14}N(n, p)$	3×10^{-3} Bq	$^{17}O(n, \alpha)$	3×10^{-5} Bq
^{81}Kr	2.1×10^{5}	$^{80}Kr(n, \gamma)$	1000 atoms	None	0
^{36}Cl	3.01×10^{5}	$^{35}Cl(n, \gamma)$	10^{-5}–10^{-3} atoms	$^{35}Cl(n, \gamma)$	5×10^{7} atoms
^{129}I	16×10^{6}	Spontaneous fission	5×10^{4} atoms	Spontaneous fission	3×10^{5} atoms

have been used to accurately determine the history of groundwater dating back from 50 to 40,000 years, respectively [22].

Carbon-14 (^{14}C) is utilized due to its half-life of 5730 years. This process facilitates the identification of the origins of dissolved inorganic carbon (DIC) within the geochemical development of groundwater and enables the tracing of its historical pathway [22]. Dissolved inorganic carbon in groundwater can originate from various isotopic sources, including soil CO_2 from respiration and organic matter decay through microbial activity, as well as dissolved carbonates resulting from the dissolution of carbonate minerals [25]. Additionally, atmospheric carbon can act as a carbon source. The δ^{13}CDIC value for rainwater in equilibrium with atmospheric CO_2 is approximately $-9‰$, while the $\delta^{13}C$ for current atmospheric CO_2 is $-8‰$ [24]. Carbon dioxide, dissolved inorganic carbon, and dissolved organic carbon extracted from living biomass exhibit 14C activities nearby the modern atmospheric level of ~100 pMC. The ^{14}C method has been extensively used to determine the history of groundwater dating back to the late Quaternary period [26].

3 Evaluation of Groundwater Aquifer Recharge, Storage, and Recovery in the MENA Region

Managed Aquifer Recharge (MAR) is considered a successful strategy for the long-term management of groundwater in MENA countries and has been studied at various sites in the region [10–16]. In the MENA region, effectively managing water resources hinges on accurately determining groundwater age and differentiating between the availability of ancient fossil water and sustainable sources of groundwater. Groundwater geochemistry and isotopes such as ^{18}O, ^{2}H, $^{87}Sr/^{86}Sr$, and ^{11}B have been employed to determine groundwater recharge pathways, evaluate historic and current geochemical processes influencing groundwater quality in arid countries [26–29], and provide insights into groundwater sources, mixing ratios, and the extent of water-rock interactions. Strontium isotopes ($^{87}Sr/^{86}Sr$) have been particularly useful in this regard [26, 29]. Boron isotopes have also been utilized to investigate the migration of pollutants in groundwater [30]. Environmental isotopes have revealed that fresh groundwater does not originate from seawater or other saline sources [26, 31]. Moreover, combining ^{11}Br and ^{81}Br isotopes can provide detailed information on processes influencing seawater intrusion and groundwater salinization in coastal aquifers [32].

3.1 Countries of the Gulf Cooperation Council (GCC)

The GCC area, part of the broader MENA region, faces significant water-related challenges. In recent decades, swift population growth and fast-paced socio-economic advancements in the GCC nations have led to a sharp increase in water consumption. From about 6 billion cubic meters (BCM) in 1980, water usage surged to roughly 30 BCM by 2010 and is anticipated to exceed 50 BCM by 2030. Despite efforts to expand desalination and wastewater recycling, groundwater continues to be the main water source. Yet, the swift depletion and diminishing of groundwater reserves have led to a decline in water quality and increased salinity in agricultural soils.

In the investigation of Kuwait's groundwater, isotopes such as deuterium (^{2}H), carbon-14 (^{14}C), and tritium (^{3}H) were utilized to determine the origin of replenishment [33]. Oxygen-18 (^{18}O) and deuterium (^{2}H) were used to investigate the origin of water in the aquifer system, while carbon-14 (^{14}C) and tritium (^{3}H) were employed to estimate the age and circulation of groundwater [33]. The findings indicated limited replenishment from recent rainfall based on carbon-14 and tritium analyses. The isotopic characteristics of groundwater and potential replenishment sources varied significantly, with the water from the Dammam formation and the Kuwait group having different ages, suggesting no replenishment from these sources. However, the similarity in isotopic composition between the Dammam aquifer and the overlying Kuwait group aquifer confirmed their hydraulic interconnection [33]. The conclusion was drawn that the elevation of groundwater levels near the study area can be traced back to heightened seepage from the Dammam formation into the groundwater, coupled with the percolation of rainwater and the spillover from adjacent disposal pits.

The use of oxygen-18 (^{18}O) has been helpful in determining the amount of groundwater replenishment in the Arabian Peninsula. Groundwater with significant recent replenishment exhibits $\delta^{18}O$ values ranging from -3.5 to 0%. Deuterium excess values in contemporary groundwater across the Arabian Peninsula largely align with the global meteoric water line (GMWL), with notable exceptions in certain local and sub-regional areas. Groundwater from various parts of the Arabian Peninsula and the eastern coastal areas of the United Arab Emirates demonstrate δ18O values between -2.0 and -3.3%, indicative of recent recharge events. In Qatar, the mean delta values for shallow fresh to brackish groundwater are δ18O -2.32% and δ2H $-$ 8.6%, reflecting the specific hydrological conditions and recharge processes in these regions [34].

In the UAE, renewable water resources are extremely limited, and rainfall is scarce and infrequent, with a yearly average precipitation of about 120 mm. Rainfall mostly occurs during a few events [35]. Additionally, there has been a significant depletion of the groundwater table in most aquifers due to an imbalance between recharge and discharge. To develop water resources in the country, numerous detention and retention dams have been constructed across the main wadis. Isotope methods, including environmental isotopes such as ^{18}O, ^{2}H, and ^{3}H, have been used to assess the increase in groundwater recharge resulting from these dams in major wadis such as Wuraiah, Bih, and Tawean. Isotopic evidence suggests that these dams are crucial for groundwater recharge. The data allow for the estimation of artificial recharge, which is calculated to account for 20–40% of the water stored by these structures [35].

3.2 Mashreq Countries

In Lebanon, geochemical parameters and stable isotopes such as oxygen-18 and deuterium are used to determine the source and mixing of groundwater in submarine aquifers. The isotopic compositions of groundwater indicate that precipitation and infiltration from surface water are the main sources of recharge [36]. Additionally, the stable isotopic ratio of sulfur isotopes ($\delta^{34}S$ and $\delta^{18}O$ in sulfate) is used to differentiate between maritime, evaporitic, and volcanic origins of dissolved sulfate in groundwater, providing insights into special chemical processes in surface and groundwater [37].

In southern Jordan, the unconfined part of the Disi sandstone aquifer has been studied using C-14 dating. The results show C-14 values ranging from 26 to 33 pMC, suggesting that this part of the aquifer either receives modern replenishment or the last replenishment occurred between 5,000 and 10,000 years ago during the Pleistocene to early Holocene period [34].

A report by the International Atomic Energy Agency (IAEA) on Syria highlights the effective use of stable isotopes, tritium, and geochemical data to monitor artificial replenishment schemes involving the injection of water into an aquifer. In areas such as the Jebel el Arab mountain region in the western Golan Heights and certain locations in the Hauran plateau, tritium values ranging from 5 to 21 TU indicate recent groundwater replenishment [34].

Carbon-14 analyses indicate contemporary groundwater recharge in the Suweida sector and the eastern regions of the Hauran plateau, which cover parts of southern Syria and northern Jordan. For instance, within the Khabour sector of Syria's Jezire, Carbon-14 dating reveals that the groundwater in the Paleogene aquifer is around 9,000 years old, while ages for the deeper Upper Cretaceous aquifer range from 6,700 to 12,000 years [34].

In southern Syria, the Ain el Fije spring water is characterized by a mix of recent recharge and the movement of older groundwater. The average time the groundwater spends in this region is roughly estimated at 50 years, based on a Carbon-14 (C-14) measurement of 62% modern carbon (pMC). In a similar vein, the As Sinn spring, located at the base of the Ansariye Mountains near the Mediterranean coast, has an estimated groundwater mean retention time of 60 years, with a C-14 level of 58 pMC. When comparing the C-14 and tritium measurements of groundwater in the aquifers of the Qalamoun high plain, an initial C-14 measurement of 68 pMC was observed, indicating a need to recalibrate the C-14 ages of the water by approximately 3,000 years [34].

3.3 North Africa Countries

In the El Dabaa region along the northwestern coast of Egypt, groundwater serves as a crucial source of fresh water for drinking, household use, and irrigation. However, in recent years, there has been a noticeable increase in the salinity of the groundwater [38]. To identify the main contributors to groundwater replenishment and the factors behind its salinization, researchers have turned to geochemical methods and the use of environmental isotopes. The investigation centered on two key aquifers: the porous oolitic Pleistocene aquifer and the fractured limestone Miocene aquifer [38]. Findings indicate that the former aquifer is primarily replenished by meteoric water, through leaching and dissolution, whereas the latter aquifer exhibits signs of seawater intrusion, attributed to excessive groundwater extraction. The assessment of the seawater mixing index (SMI) in groundwater samples provided further evidence of this seawater intrusion [38].

In the El Dabaa area, low precipitation and overuse of groundwater have led to a degradation in groundwater quality in the alluvial coastal plain aquifer downstream. This has resulted in a scarcity of annual groundwater recharge. To assess the replenishment sources, water-rock interactions, and seawater mixing for effective groundwater managing policies, isotopes such as ^{18}O, ^{2}H, $^{87}Sr/^{86}Sr$, ^{81}Br, and ^{11}B, along with groundwater geochemistry, were utilized. The study determined that effective management of groundwater withdrawal is essential to mitigate salinization and prevent further seawater intrusion into the groundwater supply [38].

Isotope analysis in Saini, Egypt, revealed that the Quaternary coastal aquifer contains mixtures of 5–13% seawater and 87–95% fresh groundwater. These findings highlight the critical need for cautious management of future groundwater extraction to avoid additional salinization. It is advised that the volume of groundwater extracted from the Quaternary coastal aquifer should not exceed the aquifer's natural average recharge rate to maintain its sustainability [38].

In Tunisia, the Kairouan plain is an important aquifer in central Tunisia, covering a large endorheic sedimentary basin of 3000 km^2. This aquifer is significant due to its groundwater storage capacity and replenishment from major wadis (Zeroud and Merguellil) floods and direct rainfall infiltration. However, the construction of the Sidi Saâd dam on Oued Zeroud to protect the plain from flooding has led to reduced natural replenishment. The overutilization of groundwater for anthropic needs has caused a stabilized depletion of the water level, despite regular dam water releases. Isotopic investigations have been carried out to assess the effectiveness of artificial replenishment. Stable isotopes (^{2}H, ^{18}O) were used to trace water movement, determine water residence time, and improve the hydrodynamic characterization of the aquifer system. Radioactive isotopes (^{14}C, ^{3}H) were utilized to assess the impact of dam water on groundwater. The stable isotopic signature of dam water is crucial for understanding its influence on groundwater [39].

In the Souss-Massa coastal plain in Morocco, the primary water source is the Souss-Massa Plio-Quaternary aquifer and several reservoirs. Isotopic data show that the key source of replenishment for the superficial aquifer is the Atlas Mountains, which receive significant rainfall (−6 to −7.5‰ $\delta^{18}O$). The isotopic composition of deuterium and oxygen-18 shows a westward gradient with more depleted values. The occurrence of enriched values in irrigated areas indicates induced recharge from irrigation return [40].

4 Groundwater Age

Isotopes are valuable tools for estimating the age, vulnerability, and sustainability of water resources. The age of groundwater can be determined based on its isotopic composition, with "old" groundwater indicating slow flow and long replenishment times, while "young" groundwater is easily renewed but more susceptible to pollution and climate change. Tritium is used to date groundwater up to approximately 60 years old, while carbon-14 can be used for water up to 40,000 years old, and krypton-81 for water that can be up to millions of years old. The Nubian Sandstone Aquifer, a significant transboundary groundwater resource for Egypt, Libya, Chad, and Sudan, has been studied using isotopes to determine groundwater age. Reliable information and monitoring tools are crucial for the sustainable management of this aquifer [39]. The use of krypton isotopes (^{81}Kr) has revealed groundwater ages ranging from nearly 200,000 to 1.5 million years in productive wells near oases in the western desert of Egypt, indicating the value of isotopes in estimating groundwater age]48].

5 Submarine Groundwater Discharge Assessment Along MENA Region

Improper management of groundwater resources can lead to various challenges, including excessive pumping costs, reduced water yield, seawater intrusion in coastal areas, and submarine groundwater discharge (SGD). To accurately assess groundwater resources and budgeting, it is essential to study hydrogeological parameters and aquifer properties more precisely. SGD refers to the flow of freshwater through coastal aquifers or the recirculation of seawater through sediments, or a combination of both [41]. SGD detection can serve as an additional water source for coastal communities and help prevent saltwater intrusion into coastal groundwater zones. Employing radon (Rn) and radium (Ra) isotopes as tracers, in conjunction with chemical and biological metrics—like temperature, electrical conductivity, salinity, dissolved oxygen, nitrates, sulfate, and biological oxygen demand—can yield insights into the origins, flow paths, exchange dynamics, and chemical properties of SGD [42]. Submarine Groundwater Discharge (SGD) can be identified through four distinct conditions, which are defined by the correlation between electrical conductivity (EC) and radon levels in water: (1) an increase in radon and a decrease in EC indicate a freshwater SGD process, (2) an increase in radon and an increase in EC indicate a saline SGD process, (3) a decrease in radon and a decrease in EC indicate fresh surface water, and (4) a decrease in radon and an increase in EC indicate a seawater intrusion process [43].

In-situ measurements of radon ^{222}Ra can be used to identify submarine groundwater discharge locations in Kuwait Bay or the Arabian Gulf [21]. Laboratory measurements of ^{223,224,226,228}Ra can help determine the amount of submarine groundwater discharge using the offshore-to-beach gradient model. The ^{224}Ra/^{228}Ra ratio can provide information about the mean residence time of water in the estuary, while the $^{228}Ra/^{226}Ra$ ratio can aid in identifying possible submarine groundwater discharge occurring through the deep carbonate layers of the Dammam Formation. Investigations into submarine groundwater discharge are essential for devising strategies for the sustainable management of coastal aquifers, keeping the interface between seawater and freshwater within coastal regions intact.

6 Groundwater Salinization Studies Using Tracer Techniques

In the Mediterranean region, the rising demand for high-quality water is met with the challenge of salinization, which threatens the expansion of water resources like groundwater. Salinization in coastal aquifers arises from seawater intrusion and water-rock interactions, two processes that are often difficult to differentiate. This increase in groundwater salinity significantly compromises water quality. Coastal areas are especially prone to salinization, which can be attributed to a variety of causes such as the subsurface migration of seawater, the ascent of ancient seawater trapped during sediment deposition or previous high sea levels, leakage of highly saline water from evaporation in tidal lagoons, and the reintroduction of saline wastewater and runoff from agricultural practices [44].

Geochemical analysis of main and secondary elements, along with nuclear tracer techniques, can be used to determine the sources of salt in groundwater [29]. Analyzing concentrations of elements like bromine (Br) and ratios of chloride can aid in determining the source of salinity, especially in cases of seawater intrusion [45–47]. The concentration of total dissolved solids (TDS) and the levels of various ions can divide groundwater into different groups based on their salinity levels. Salinity levels can be influenced by processes such as cation exchange, mineral dissolution or precipitation, anthropogenic contaminants, saltwater intrusion, inverse cation exchange, and evaporation.

Isotopes play a crucial role in identifying the sources of groundwater salinity and understanding the rates of groundwater flow. Each source of salinity has a unique chemical and isotopic composition. When salt concentrations increase (salinization), environmental isotopes like deuterium (^{2}H) and oxygen-18 (^{18}O) can be effectively used to determine the sources of saline groundwater [48, 49]. Environmental isotopes have become important tools for studying the origin of declining groundwater quality. The phenomenon of seawater intrusion, for example, can be investigated using isotopes such as ^{11}Br and ^{81}Br, which provide detailed information on the processes affecting groundwater salinization in coastal aquifers. In the Dahab basin in Egypt, ^{81}Br isotopic values in groundwater ranged from 0.24 h to +2.15 h, indicating variations in seawater intrusion.

The environmental isotopes ^{2}H and ^{18}O have been used to study the geochemical differences in high-salinity groundwater in Baghdad's central business district in

Iraq. The isotopic analysis revealed multiple sources of salinity, including older water and dissolved water from Holocene and Pleistocene deposits, as well as contamination from human activities [50]. The correlation between stable oxygen-18 isotopes and electrical conductivity can provide insights into the origin of increased salt concentration in groundwater. Changes in stable isotopic composition [46, 47], are associated with mixing and/or evaporation processes. In some cases, groundwater samples with similar oxygen-18 levels have similar salinity levels, indicating the influence of dissolved salts, sewage mixing, and water tank contamination. Dissolution and mixing processes are significant factors in determining the origin of salts in groundwater. Additionally, anthropogenic activities, such as mixing with sewage water [50], can contribute to the salinity of groundwater. In conclusion, various methods, including geochemical analysis and tracer techniques utilizing environmental isotopes, are valuable for studying groundwater salinization. These techniques help identify the sources of salinity, understand the underlying processes, and guide management strategies to ensure the sustainable use of water resources.

7 Conclusions

Groundwater, being a scarce and vulnerable strategic natural resource in the MENA region, requires vigilant monitoring, efficient management, conservation, and protection to ensure its long-term sustainability. The region faces challenges including overexploitation, pollution, and climate change, leading to the deterioration of both quantity and quality of groundwater resources. This degradation not only affects the ecosystems but also has significant economic impacts. To address these issues, innovative approaches and advanced technologies are crucial for efficient governance. Isotopic techniques have emerged as invaluable tools for identifying and quantifying groundwater recharge in arid and semi-arid climatic conditions as well as assessing and monitoring the sea water intrusion cause by the sea level rise due to impact of climate change on the coastal areas.

In the Arabian Peninsula, the use of Oxygen-18 has proven to be effective in identifying the quantities of groundwater recharge, while C-14 has been instrumental in determining the age of recharge in southern Jordan. Isotopic techniques also play a crucial role in pinpointing seawater intrusion, which is amplified by the impact of climate change-induced sea level rise. For instance, the Bromine isotope has been employed to detect seawater intrusion in seaboard aquifers in Egypt, aiding in the control of water and soil salinity in the agriculturally significant Northern Delta region. Additionally, the Krypton isotope (Kr-81) has been utilized for managing transboundary aquifer systems, specifically in the Nubian Sandstone Aquifer, and assessing groundwater extracted from wells near oases in the western desert of Egypt.

8 Recommendations

- Regular monitoring of groundwater resources in the MENA region is of paramount importance to ensure sustainable utilization and to implement necessary measures for improving water quality.
- Building the capacity of relevant stakeholders involved in groundwater management and fostering the development of technical, legal, and institutional frameworks are crucial for enabling the effective use of nuclear techniques in groundwater monitoring and management.
- Effective partnerships between different responsible entities are essential for leveraging the data and knowledge generated by academia and research bodies to enhance groundwater governance and decision-making processes.

References

1. UNDP (United Nations Development Programme))2013(Water governance in the Arab region, managing scarcity and securing the future. UNDP, 182 pages. https://www.undp.org/arab-states/publications/water-governance-arab-region
2. ACSAD (Arab center for the studies of arid zones and dry Lands) (2009) Water resources management in the Arab region, second bulletin. Published in cooperation with GTZ and NCRS in Lebanon
3. Ruta G (2005) Deep wells and shallow savings: the economic aspect of groundwater depletion in MENA countries. In: Background paper to making the most scarcity: accountability for better water management results in the Middle East and North Africa. World Bank, Washington, DC
4. Ghoudi R (2017) Groundwater governance in Tunisia. Expert group meeting on groundwater governance, Bahrain
5. El Zaher O (2017). Groundwater governance in Egypt. Expert group meeting, Bahrain
6. Baleid KM (2017) Water resources and irrigation in Yemen. Expert group meeting on groundwater governance, Bahrain
7. Khater AR (2010) Regional technical report on the impacts of climate change on groundwater in the Arab region. Technical document published by UNESCO Cairo office
8. Bohannon J (2010) The Nile Delta's Sinking future. Science 327:1444–1447. https://doi.org/10.1126/science.327.5972.1444
9. El Raey (2010) Impacts and implications of climate change for the costal zones of Egypt. In: Mechel D, Pandya A (eds) The henry L. Simon Center, Washington DC, p 31050
10. El-Rawy M, Zlotnik VA, Al-Raggad M, Al-Maktoumi A, Kacimov A, Abdalla O(2016) Conjunctive use of groundwater and surface water resources with aquifer recharge by treated wastewater: evaluation of management scenarios in the Zarqa River basin, Jordan. Environ Earth Sci 75:1–21. https://doi.org/10.1007/s12665-016-5946-1
11. Al-Maktoumi A, El-Rawy M, Zekri S (2016) Management options for a multipurpose coastal aquifer in Oman. Arab J Geosci 9:636. https://doi.org/10.1007/s12517-016-2661-x
12. El-Rawy M, Al-Maktoumi A, Zekri S, Abdalla O, Al-Abri R (2019) Hydrological and economic feasibility of mitigating a stressed coastal aquifer using managed aquifer recharge: a case study of Jamma aquifer, Oman. J Arid Land 11:148–159. https://doi.org/10.1007/s40333-019-0093-7
13. Abdelrady A, Sharma S, Sefelnasr A, El-Rawy M, Kennedy M (2020) Analysis of the performance of bank filtration for water supply in arid climates: case study in Egypt. Water 12(6):1816

14. Al-Maktoumi A, Zekri S, El-Rawy M, Abdalla O, Al-Abri R, Triki CMR, Bazargan-Lari MR (2020) Aquifer storage and recovery, and managed aquifer recharge of reclaimed water for management of coastal aquifers. Desalin Water Treat 1(176):67–77. https://doi.org/10.5004/dwt.2020.25499
15. Ajjur SB, Baalousha HM (2021) A review on implementing managed aquifer recharge in the Middle East and North Africa region: methods, progress and challenges. Water Int 46(4):578–604. https://doi.org/10.1080/02508060.2021.1889192
16. Sherif M, Sefelnasr A, Al Rashed M, Alshamsi D, Zaidi FK, Alghafli K, Baig F, Al-Turbak A, Alfaifi H, Loni OA, Ahamed MB (2023) A review of managed aquifer recharge potential in the Middle East and North Africa Region with examples from the Kingdom of Saudi Arabia and the United Arab Emirates. Water 15(4):742. https://doi.org/10.3390/w15040742
17. Dillon P (2009) Water recycling via managed aquifer recharge in Australia. Boletín Geológicoy Minero 120(2):121–130
18. Dillon P, Arshad M (2016) Managed aquifer recharge in integrated water resource management. In: Integrated groundwater management: concepts, approaches and challenges, pp 435–452
19. Page D, Bekele E, Vanderzalm J, Sidhu J (2018) Managed aquifer recharge (MAR) in sustainable urban water management. Water 10(3):239. https://doi.org/10.3390/w10030239/
20. Zhang H, Xu Y, Kanyerere T (2020) A review of the managed aquifer recharge: historical development, current situation and perspectives. Phys Chem Earth, Parts A/B/C 118:102887
21. Kumar US, Hadi K (2018) Futuristic isotope hydrology in the Gulf region. Appl Water Sci 8:25. https://doi.org/10.1007/s13201-018-0647-4
22. Aggarwal P, Froehlich K, Gonfiantini R, Gat JR (2009) Environmental isotopes in groundwater studies. In: Aggarwal PK, Froehlich K, Kulkarni Km (eds). EOLSS Publications, 584 pages
23. Lapworth DJ, MacDonald AM, Krishan G, Rao MS, Goody DC, Darling WG (2015) Groundwater recharge and age-depth profiles of intensively exploited groundwater resources in northwest India. Geophys Res Lett 42. https://doi.org/10.1002/2015GL065798
24. Kumar S (2019) Environmental isotopes in groundwater applications. In: Sikdar P (ed) Groundwater development and management. Springer, Cham. https://doi.org/10.1007/978-3-319-75115-3_4
25. Clark I, Fritz P (1997) Environmental isotopes in hydrogeology. Lewis Publishers, Boca Raton, FL, 328 p
26. Kalin RM (2000) Radiocarbon dating of groundwater systems. In: Environmental tracers in subsurface hydrology, Chapter: 4. Kluwer Academic Publishers, pp 111–144. https://doi.org/10.1007/978-1-4615-4557-6_4
27. Eissa MA, Thomas JM, Pohll G, Shouakar-Stash O, Hershey RL, Dawoud M (2016) Groundwater recharge and salinization in the arid coastal plain aquifer of the Wadi Watir delta, Sinai, Egypt. Appl Geochem 71:48–62
28. Issar A, Gilad D (1982) Groundwater flow system in the arid crystalline province of southern Sinai. Hydrol Sci J 27:309–325
29. Edmunds WM (1995) Characterization of groundwaters in semi-arid and arid zones using minor elements. In: Nash HG, McCall J (eds) Groundwater quality. Chapman & Hall: London, UK, pp 19–30
30. Quast KW, Lansey K, Arnold R, Bassett RL, Rincon M (2006) Boron isotopes as an artificial tracer. Groundwater 44:453–466. https://doi.org/10.1111/j.1745-6584.2005.00186.x
31. Du Y, Ma Y, Chen L, Shan H, Xiao C, Lu Y, Liu C, Cai H (2015) Genesis of salinized groundwater in quaternary aquifer system of the coastal plain, Laizhou Bay, China: geochemical evidence, especially from bromine stable isotope. Appl Geochem 59:155–165. https://doi.org/10.3390/w15213788
32. Stotler RL, Frape SK, Shouakar-Stash O (2010) An isotopic survey of d81Br and d 37Cl of dissolved halides in the Canadian and fennos Canadian shields. Chem Geol 274:38–55. https://doi.org/10.1016/j.chemgeo.2010.03.014
33. Bhandary H, Al-Fahad K, Al-Senafy M, Al-Khalid A (2012) Usage of environmental isotopes in characterizing groundwater recharge sources. WIT Trans Ecol Environ 164. www.witpress.com. https://doi.org/10.2495/WP120191. ISSN 1743-3541 (WIT Press)

34. Wagner W (2011) Groundwater in the Arab Middle East. Science. Springer Science & Business Media, Heidelberg, Dordrecht, London, New York, 443 pages. https://doi.org/10.1007/978-3-642-19351-4
35. Ekaabi AS (2007) Application of isotope hydrology for the assessment of artificial ground water recharge in some areas of UAE. In: Proceedings of an international symposium on advances in isotope hydrology and its role in sustainable water resources management (HIS-2007) organized by international atomic energy agency in Vienna
36. Saad Z, Kazpard V (2007) Use of environmental isotopes to evaluate the sources of submarine freshwater in the southern shoreline in Lebanon. In: Proceedings of an international symposium on advances in isotope hydrology and its role in sustainable water resources management (HIS-2007) organized by international atomic energy agency in Vienna
37. Krouse HR (1980) Sulphur isotopes in our environment. In: Fritz P, Fontes JC (eds) Handbook of environmental isotope geochemistry vol I, The Terrestrial Environment. Elsevier, Amsterdam, pp 435–471
38. Eissa MA, Shawky H, Samy A, Khalil MMH, El Malky M (2018) Geochemical and isotopic evidence of groundwater salinization Processes in El Dabaa Area, Northwestern Coast, Egypt. Geosciences 8:392
39. IAEA (2010) Newsletter of the Isotope Hydrology Section. Issue No. 26. International Atomic Energy Authority, Vienna
40. Bouchaoua L, Qurtobi M, Hsissoua Y, Ibn Majahb M, Marahb H (2009) Isotopic techniques as a tool to investigate salinization problems in the Souss Massa coastal plain (South West Morocco). In: Origin of salinity and impacts on fresh groundwater resources: Optimization of Isotopic techniques (Results of a 2000–2004 Coordinated Research Project), IAEA, Water Resource Programme
41. Mook WG (2001) Environmental isotopes in the hydrological cycle: principles and applications. United Nations Educational Scientific and Cultural Organization (UNESCO), Paris
42. IAEA, WISER (2008) Water isotope system for data analysis, visualization, and electronic retrieval. WISER Version 0.7. https://websso.iaea.org. Accessed 1 May 2013
43. Krishan G, Someshwar Rao M, Kumar CP, Kumar S, Rao MRA (2015) A study on the identification of submarine groundwater discharge in the northern east coast of India. Aquatic Procedia 4:3–10
44. Ahmed MA, Abdel Samie SG, Badawy HA (2013) Factors controlling mechanisms of groundwater salinization and hydrogeochemical processes in the quaternary aquifer of the Eastern Nile Delta, Egypt. Environ Earth Sci 68:369–394
45. Rao U, Hollocher K, Sherman J, Eisele I, Frunzi MN, Swatkoski SJ, Hammons AL (2005) The use of 36Cl and chloride/bromide ratios in discerning salinity sources and fluid mixing patterns: a case study at Saratoga Springs. Chem Geol 222:94–111
46. El Moujabber M, Bou Samra B, Darwish T, Atallah T (2006) Comparison of different indicators for groundwater contamination by seawater intrusion on the Lebanese coast. Water Resour Manag 20:161–180
47. Park Y, Lee JY, Kim JH, Song SH (2012) National scale evaluation of groundwater chemistry in Korea coastal aquifers: evidences of seawater intrusion. Environ Earth Sci 66:707–718. https://doi.org/10.1007/s12665-011-1278-3, 589–605
48. Martos FS, Bosch AP, Sanchez LM, Izquierdo AV (2002) Identification of the origin of salinization in groundwater using minor ions (lower Andarax, Southeast Spain). J S Total Env 297:43–58
49. Boluda-Botella N, Gomis-Yagües V, Ruiz-Bevia F (2008) Influence of transport parameters and chemical properties of the sediment in experiments to measure active transport in seawater intrusion. J Hydrol 357:29–41
50. Al-Charideh AR (2010) Environmental isotope study of groundwater discharge from the large karst springs in the west Syria. Environ Earth Sci 23(2). https://doi.org/10.1007/s12665-010-0660-x

Rainwater Harvesting for Managed Aquifer Recharge and Flood Mitigation in the MENA Region

Mustafa El-Rawy, Mohamed Wahba, and Heba Fathi

Abstract Water security is a pressing global concern, with particular significance for the Middle East and North Africa (MENA) region, which faces acute water scarcity challenges. MENA countries are among the most water-stressed in the world due to a combination of factors, including rapid population growth, climate change-induced droughts, and over-extraction of groundwater resources. This chapter provides a comprehensive overview of Rainwater Harvesting (RWH)) for Managed Groundwater Recharge (MAR) and Flood Mitigation (FM). The chapter aims to investigate the various rainwater harvesting techniques, both traditional and modern, evaluating their applicability in diverse contexts and their role in mitigating flood risks. Additionally, the chapter will evaluate different methods of managed groundwater recharge, emphasizing their effectiveness in replenishing groundwater reserves and reducing the impact of floods. Finally, the chapter aspires to contribute to the understanding and advancement of sustainable water management practices and flood mitigation strategies. RWH techniques play a crucial role in supporting MAR and FM strategies. MAR involves the controlled replenishment of underground aquifers with rainwater, and these techniques serve as a primary source of this valuable resource. RWH systems, such as rooftop collection systems and surface runoff management, capture and channel rainwater for storage and subsequent infiltration into aquifers. This process not only helps restore groundwater levels but also improves water quality as the collected rainwater is usually free from contaminants. Additionally, by effectively

M. El-Rawy (✉)
Civil Engineering Department, Faculty of Engineering, Minia University, Minia 61111, Egypt
e-mail: mustafa.elrawy@mu.edu.eg

Civil Engineering Department, College of Engineering, Shaqra University, Dawadmi 11911, Saudi Arabia

M. Wahba
Civil Engineering Department, Faculty of Engineering, Mansoura University, Mansoura 35516, Egypt
e-mail: m_wahba@mans.edu.eg

H. Fathi
College of Design and Architecture, Jazan University, Jazan 45142, Saudi Arabia
e-mail: hamansour@jazanu.edu.sa

M. El-Rawy and A. Negm (eds.), *Managed Aquifer Recharge in MENA Countries*, Earth and Environmental Sciences Library,
https://doi.org/10.1007/978-3-031-58764-1_3

managing excess rainwater during heavy downpours, these techniques contribute to flood mitigation efforts, reducing the risk of flooding in urban and rural areas. The integration of rainfall harvesting into MAR and flood mitigation strategies offers a sustainable and adaptable solution for addressing water scarcity and managing the impacts of extreme weather events.

Keywords RWH · Flood mitigation · Artificial recharge wells · Extreme rainfall · Water resources management · MENA

1 Introduction

Throughout history, there has been a recorded human tendency to relocate to more fitting environments in response to climate change and periods of drought [1, 2]. The rainwater harvesting systems are historical methodologies that have fulfilled the requirements of communities over numerous generations, signifying their evident sustainability. This approach involves the process of stimulating, gathering, stockpiling, and preserving local runoff for subsequent beneficial utilization. Presently, the prevalence of rainwater harvesting systems has dwindled significantly due to the consolidation of water reserves, facilitated by amplified groundwater extraction capabilities and the rerouting of surface water across extensive distances through river damming and channeling. This pattern has demonstrated its lack of sustainability in numerous areas, notably affecting tropical regions characterized by substantial populations of small-scale farmers who consistently face water accessibility challenges amidst the process of modernization [3].

Rainwater collection and storage played a pivotal role in the prosperity of ancient societies, such as those in the Middle East, Southeast Asia, and Central America [4, 5]. Residents of these areas managed to not only survive but also thrive in climates ranging from dry-wet to semi-arid and arid, which are among the most demanding environmental conditions faced by humans. These regions and their civilizations confronted various scales of climate change events during the Late Holocene period, as documented by Wanner et al. [6]. Moreover, in recent years, there has been a heightened global interest in rainwater harvesting across both developing nations and advanced economies, including several EU Member States [7]. The efficacy of RWH systems significantly hinges on their technical blueprint and the identification of optimal locations. Pinpointing suitable sites for diverse RWH technologies across extensive regions presents a considerable challenge. Consequently, numerous scholars have formulated and put into practice diverse methodologies and criteria to pinpoint appropriate sites and methodologies for effective rainwater harvesting.

In Africa, particularly in Northern Africa, there exists an extensive historical background in water harvesting (WH), wherein the method was frequently tailored to suit the landscape of individual regions. Recognized historically as the primary food source for the Roman Empire, Libya, in particular, utilized runoff irrigation extensively to cultivate various crops like barley, wheat, olives, grapes, figs, and dates within its dry terrain. This particular WH technique also facilitated the rearing of sheep, pigs, and cattle. This agricultural framework endured for more than 400 years, supporting a sizable sedentary populace, often affluent, that produced ample harvests to yield surplus crops [8]. On the other hand, the adoption of Rainwater Harvesting (RWH) systems in European nations exhibits diverse patterns. Certain Western European countries have embraced RWH technologies to preserve urban water reservoirs. In the United Kingdom, there exists a historical practice where individuals traditionally gather and reserve rainwater for domestic purposes, such as laundry, dishwashing, and various cleaning activities [9].

The Middle East and North Africa (MENA) region stands out as the most contentious area globally in terms of water resources. According to the United Nations criteria, a nation faces acute water scarcity when the annual available water per person falls below 500 m^3 [10]. Projections suggest that within the next decade, the average available water per capita in MENA will decrease further to 460 m^3 annually [10].

In the Middle East and North Africa (MENA) region, traditional methods of water collection persist across all nations. According to specialists from the Food and Agriculture Organization (FAO) specializing in soil and water conservation, there exists a pressing need to enhance the efficacy of these traditional approaches. Regrettably, the implementation of systems tested under diverse climatic, soil, land ownership, and socio-economic circumstances often encounters resistance from the intended beneficiary groups.

The principal challenges and limitations obstructing the assimilation of water harvesting into agricultural practices within MENA countries according to Nasr [11] as follows:

- Inadequate technology aligned with the specific demands of the country, region, or area.
- Limited acceptance, motivation, and engagement among potential beneficiaries.
- Scarcity of comprehensive hydrological data and information essential for confident planning, design, and execution of water harvesting initiatives.
- Inadequate consideration of social and economic aspects, including land ownership, unemployment, and the sustainable management of water harvesting systems.
- Ineffectiveness in engaging national research institutions and extension services.
- Insufficient institutional frameworks, beneficiary associations (such as cooperatives or organizations), and governmental training initiatives for farmers, pastoralists, and extension personnel.
- Lack of a sustained, long-term governmental policy supporting water harvesting endeavors.

The main objectives of this chapter are to provide a comprehensive overview of Rainwater Harvesting (RWH) for Managed Groundwater Recharge (MAR) and Flood Mitigation (FM). The chapter aims to investigate the various rainwater harvesting techniques, both traditional and modern, evaluating their applicability in diverse contexts and their role in mitigating flood risks. Additionally, the chapter will evaluate different methods of managed groundwater recharge, emphasizing their effectiveness in replenishing groundwater reserves and reducing the impact of floods. Finally, the chapter aspires to contribute to the understanding and advancement of sustainable water management practices and flood mitigation strategies.

2 Rainwater Harvesting Techniques

Rainwater harvesting constitutes a technique utilized to stimulate, gather, retain, and preserve regional surface runoff primarily for agricultural purposes in areas characterized by arid or semi-arid climatic conditions [12]. Meanwhile, arid and semi-arid (ASA) regions across the globe exhibit climatic conditions typified by scant precipitation occurring predominantly within brief periods of rainy seasons [13].

2.1 Traditional Rainwater Harvesting

2.1.1 Qanat System

It is one of the ancient and enduring water conveyance methods. This method has its origins potentially spanning over 3000 years ago [14]. These systems utilized subterranean channels, excavated to transport water primarily through gravitational force, covering distances exceeding 100 km from mountainous regions to cultivated lowlands [15, 16]. Although, the expansive water conveyance network covered extensive regions, the primary predicament facing qanats involves their swift abandonment and subsequent irregular maintenance due to the departure of younger generations from agricultural pursuits in favor of alternative livelihoods. Besides socio-economic factors, environmental conditions such as climate shifts, excessive usage, and seismic activity contribute to the abandonment of qanats. Nonetheless, despite numerous hurdles, qanats are presently being reconsidered as a viable avenue for societal advancement and conflict resolution in locations where the system endures, and favorable circumstances prevail [17].

Recent endeavors aimed at restoring qanats have showcased the significance of collaborative engagement with rural communities, thereby fostering community advancement and conflict mitigation [10]. Moreover, the restoration initiatives have underscored the dual benefits of conserving tangible and intangible cultural heritage, preserving traditional knowledge, and impeding the trend of urbanization and migration from rural to urban locales.

(a) (b)

Fig. 1 Traditional rainwater harvesting by water collation using terraced elevation; **a** from Yemen [18], **b** from South America [19]

2.1.2 Water Collection

This method involves the containment of rainfall within a specified region to hinder overall runoff, thereby extending the duration of water infiltration (see Fig. 1). Water collection techniques encompass various methods such as terraces, debris barriers, reservoirs, fog capture, and specifically, the collection of rainwater from rooftops. Historically, these methodologies primarily served the purpose of facilitating irrigation processes.

2.1.3 Rooftop Systems

In accordance with their nomenclature, rooftop and courtyard mechanisms collect and amass rainwater derived from various structures such as greenhouses, sizable edifices, residential dwellings, courtyards, and similarly robust installations. The collected runoff is often subjected to purification processes within settling basins before storage. In rural regions devoid of centralized water provisions, this decentralized method predominantly caters to domestic and potable water requirements [20]. Particularly in arid landscapes, this cost-effective approach efficiently supplies water for both human and animal needs. Presently, the utilization of rooftop systems for agricultural purposes has become increasingly prevalent (see Fig. 2).

2.2 Modern Rainwater Harvesting Systems

Innovation in the design of RWH systems is continuously evolving on a global scale. This encompasses a spectrum of approaches, from partitioning storage through interconnected modular systems and collapsible tanks [21]. to systems utilizing gutter-based collection [22], alongside other advanced, energy-efficient methodologies [23].

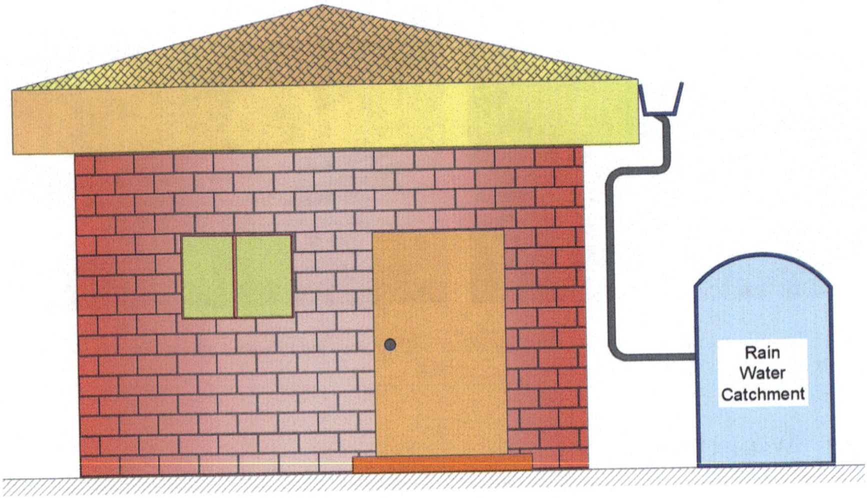

Fig. 2 Traditional rainfall harvesting by rooftop system

These adaptations are tailored to align with the specific demands and constraints of diverse environments and circumstances.

Recent studies have explored the integration of paired storage units within RWH systems [24], wherein distinct tank modules serve specific purposes of stormwater detention and retention storage. The retention component is calibrated to fulfill user needs, while the detention section, typically situated at the tank's upper portion, functions as a temporary reservoir for managing runoff. These two storage compartments can be linked by a small opening, enabling gradual drainage from the detention segment, facilitating space within the tank before subsequent rainfall [25].

RWH involves the accumulation of rainwater from impermeable surfaces, commonly rooftops, for non-potable purposes. This stored water is typically utilized for outdoor irrigation or non-potable indoor uses such as toilet flushing. An illustrative depiction of a modified domestic RWH system is presented in Fig. 3. Its constituents generally encompass a roof catchment, a filter designed to eliminate the initial portion of roof runoff, a storage tank, and a pump responsible for meeting system demands. In instances where there exists adequate hydraulic force to fulfill requirements, the use of a pump may be unnecessary. While the act of collecting rainwater serves as a water conservation measure, it might also yield benefits in terms of enhancing water quality by curbing runoff.

From another perspective, a basic framework involves utilizing the rooftop as a foundation for installing collection pipelines that channel rainwater into ground-level receptacles for immediate utilization or recharging into groundwater reserves. Figure 4 shows a diagrammatic approach to collecting rainwater from a modern rooftop system.

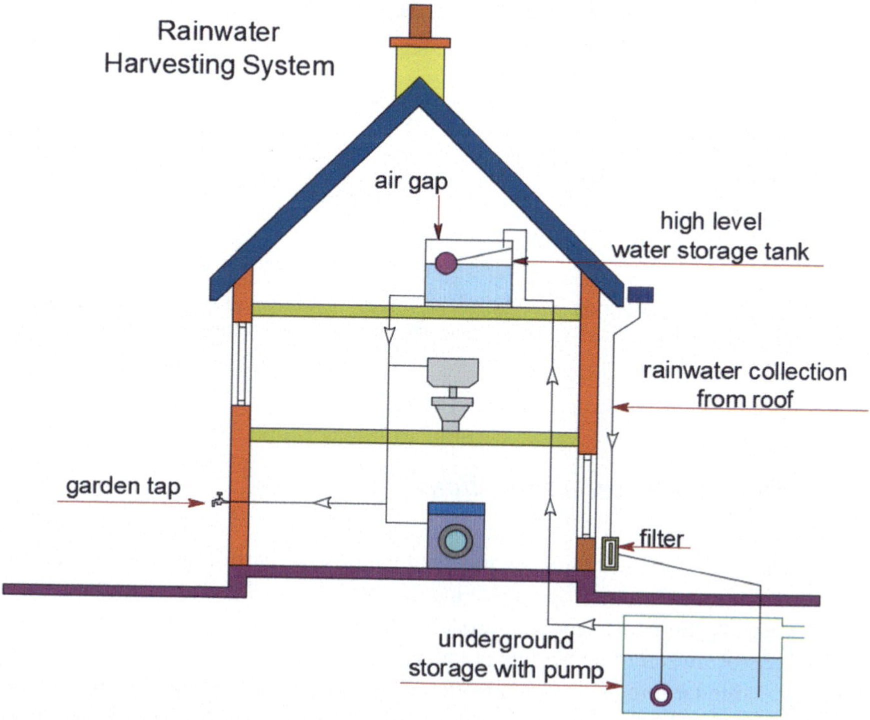

Fig. 3 Residential rainwater collection. Modified after [26]

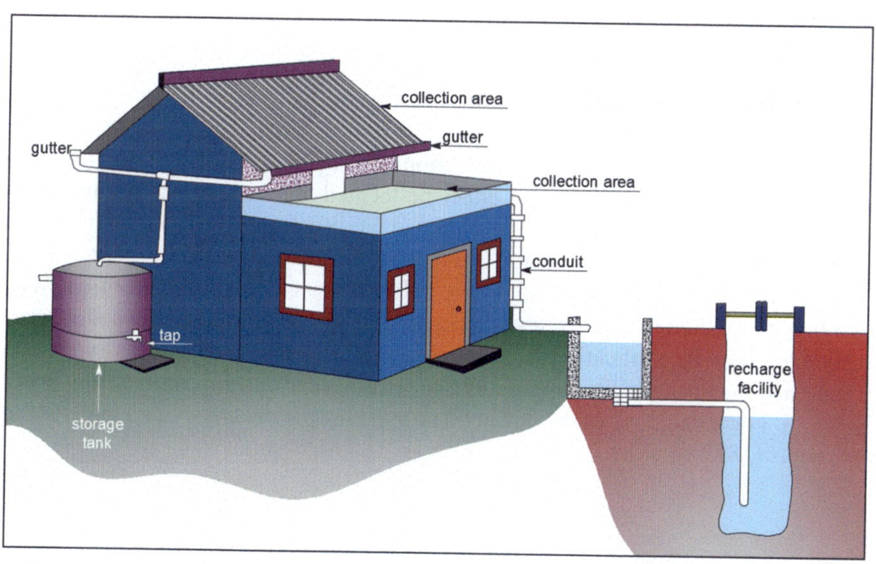

Fig. 4 Schematic method for gathering rainwater from a new rooftop setup

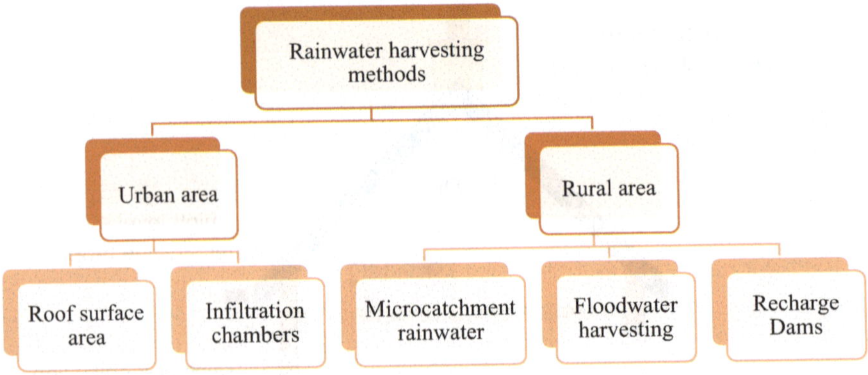

Fig. 5 Applications of RWH in urban and rural areas

2.3 *Urban and Rural Applications*

Sustainable urban rainwater management safeguards the hydrological roles of rainfall within city environments by preserving the integration of rainwater with groundwater and opting for eco-friendly approaches to collect rainwater. By implementing sustainable stormwater management practices, adverse impacts of intense rainfall are mitigated, safeguarding both groundwater and surface water, lessening the occurrence of floods and associated risks, and facilitating the availability of alternative water resources [27]. Figure 5 illustrates some applications of harvesting rainwater in urban and rural areas.

2.3.1 Roof Surface Area

Rainwater harvesting involves the collection of precipitation from building roofs utilizing gutter systems, which can either be located above or below ground. This gathered rainwater is then directed to a storage facility designed for rainwater accumulation. Subsequently, this stored water can be utilized within or outside the building, either treated or untreated. Typically, rainwater collected from rooftops is employed for domestic purposes, and after appropriate treatment, it can also serve as potable water. Its applications span various domestic uses such as laundry, garden irrigation, toilet flushing, fire suppression, pool filling, car washing, and cooling tower operation. For potable purposes, it can be utilized in specific areas like kitchens and bathrooms [28].

2.3.2 Infiltration Chambers

Infiltration refers to the downward movement of rainfall or irrigation water through the soil profile driven by gravitational forces. Infiltration tanks, a method for rainwater drainage, are employed to offer pollution control, mitigate flood risks, and aid in replenishing groundwater resources as an alternative water supply (see Fig. 6). These applications have historically served flood protection and water quality enhancement purposes. The advantages of rainwater infiltration applications include diminishing the volume of rainwater directed to drainage systems, obviating the necessity for extensive pipe networks, mitigating adverse effects associated with impermeable surfaces, contributing to the natural water cycle's completion, preserving and augmenting underground water reserves over the long term, and lessening the hydraulic load entering sewage lines. This method also aims to curtail expenses related to rainwater drainage infrastructure, decreasing the volume of rainwater conveyed to wastewater treatment plants, thereby lowering treatment costs. This method has been utilized recently in urban area by using underground tanks as alleviation method to mitigate the impact of flash flood [29].

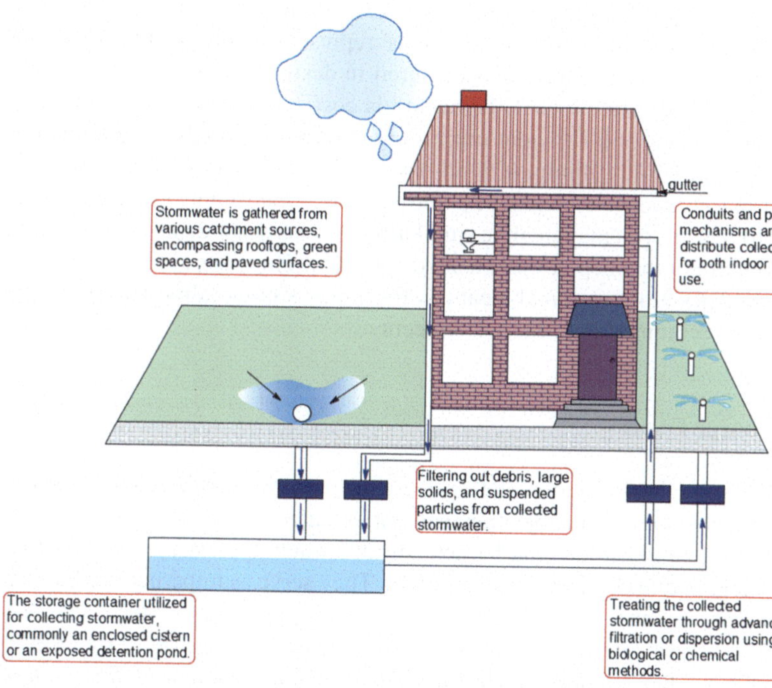

Fig. 6 Graphical framework of using the infiltration chamber for RWH

2.3.3 Microcathcment Rainwater

Micro-catchment rainwater harvesting (MIWH) systems utilize small-scale runoff catchments typically located at the household or farm level. In this approach, runoff travels as a sheet downstream toward a storage facility, intended for later utilization across different applications. Prominent variations of MIWH encompass Household systems such as rooftops and cisterns, along with Farm and Landscape systems comprising contour ridges, bunds, small runoff basins, and strips [30].

2.3.4 Floodwater Harvesting

Rainwater harvesting has been shown to reduce peak flows and mitigate flooding-related damages [30, 31]. Flood control methods involve the creation of basins in lower-lying areas to gather runoff, thereby enabling the removal of accumulated runoff water as rainfall intensity increases [32, 33]. This strategy could potentially convert otherwise destructive stormwaters into a valuable agricultural resource [34]. Various engineering-based flood mitigation solutions, such as constructing new dykes, diversions, and flood retention zones, are employed to alleviate flooding impacts [35]. Additionally, infiltration ponds represent a sustainable rainwater harvesting approach, wherein runoff is captured in designated areas and gradually seeps into the groundwater table. This method is cost-effective and poses minimal environmental impact [36]. Efficient utilization of rainwater holds the potential to augment water availability, diminish flood risks, and enhance agricultural practices, urban green spaces, and ecosystems [37]. Furthermore, identifying the regions' most susceptible to flash floods may involve simulating the necessary depth of runoff [38, 39] or creating maps that depict flood hazards [40–45]. This phase holds significant importance in determining optimal locations for implementing rainwater harvesting systems where accumulation of rainwater occurs.

2.3.5 Recharge Dams

Recharge dams, also known as percolation dams, are purposefully designed structures aimed at enhancing the sustainable management of water resources. These dams are strategically located to capture and harness surface water, typically in regions with porous soils and strong connections to aquifers. They serve as impermeable barriers built into stream channels, creating surface reservoirs upstream of the dam [46]. Recharge dams operate in two primary ways: first, by increasing surface water infiltration behind the dam, as depicted in Fig. 7; and second, by controlled release of water through the dam, allowing downstream infiltration through the riverbed [47]. Their core objective is to capture surface water during periods of heavy precipitation, temporarily storing it behind the dam, and gradually allowing it to seep into the

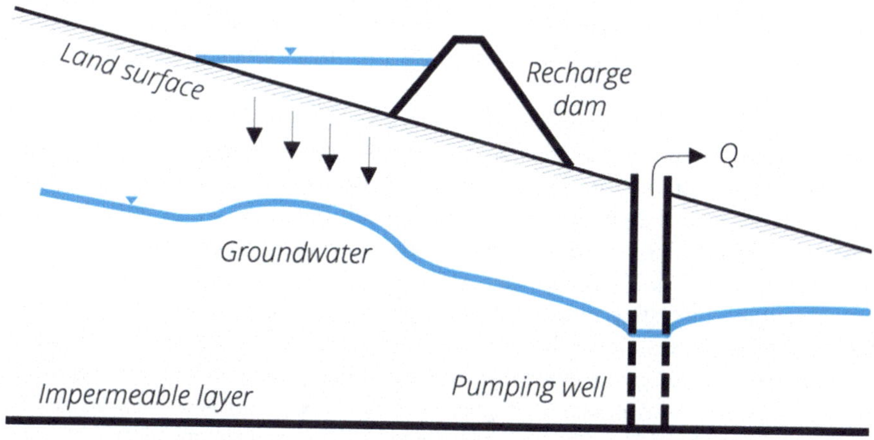

Fig. 7 Recharge dams (https://www.inowas.com/mar-methods/recharge-dams/)

soil. As this stored water percolates through the ground, it replenishes vital groundwater aquifers. Recharge dams are instrumental in ensuring a consistent and dependable supply of groundwater for a wide range of applications, including drinking water, agriculture, and industrial use and flood control purposes. Moreover, they play a critical role in mitigating the consequences of droughts and contribute to ecosystem preservation by maintaining baseflow in rivers and wetlands [48]. These structures play a crucial role in MAR and are especially important in arid and semi-arid regions where groundwater resources are vital for supporting agriculture and fulfilling domestic water requirements. Additionally, they serve as effective measures for replenishing aquifers in arid areas and mitigating seawater intrusion along coastal regions [49, 50].

3 Managed Aquifer Recharge Methods

When water supplies are not adequately managed, aquifers can become exhausted, resulting in water shortages and dryness [51]. By 2050, surface water flow in North African nations is expected to decrease by 20–30%, mostly as a result of rising temperatures and decreased precipitation, as predicted by climate models that forecast a hotter, drier, and less predictable climate. Overall, a 5–15% decrease in groundwater recharge is anticipated [52].

For groundwater that is 25–100 years old, the worldwide storage capacity of groundwater is estimated to be between 0.8 and 1.9 million km^3. In comparison, the total surface water storage in lakes and dams is two orders of magnitude smaller, at 12,900 km^3 [53], which demonstrates the importance of groundwater storage. But, the future threat points to a potential decline in groundwater recharge and the possibility of increased rainfall variability leading to more frequent and prolonged low discharge periods. As a result of rising sea levels brought due to global warming, seawater intrusion will accelerate and have an impact on the fresh groundwater resources in coastal aquifers [54].

The majority of MENA nations, especially Saudi Arabia, Jordan, Morocco, and Tunisia, have extraordinarily high sensitivities to any change in water resources, according to several findings. Fortunately, all scenarios that were simulated showed that groundwater recharge in these countries won't begin to significantly decrease until the 2050s [55]. Considering the predicted rise in more precipitation occurrences in the region, practical MAR approaches, if correctly deployed within the context of current rainfall-harvesting infrastructure, might ameliorate these issues [52]. MAR, which is a group of techniques that are being used more and more to protect, preserve, and improve stressed groundwater systems by utilizing the much larger volume of water storage that is available underground as opposed to surface water storage in lakes and dams [56, 57]. MAR can contribute to the enhancement of groundwater resources and serve as a critical factor in promoting the sustainable growth of irrigated agriculture in arid and semi-arid river basin regions. This can be achieved through the conjunctive use of both groundwater and surface water resources with aquifer recharge using treated wastewater [58]. MAR can be used to build up a strategic water resource exclusively for use during drought years, or it can be utilized to bridge the gap between irrigation in the summer and winter water supply.

The various MAR techniques that are currently being used are as follows: soil aquifer treatment (SAT), aquifer storage and recovery (ASR), aquifer storage transfer and recovery (ASTR), percolation pond, infiltration galleries, check dam, river bank filtration (RBF), rain water harvesting (RWH), dry wells, dune filtration, underground dams, and sand dams. Taking into account that aquifer types (confined, semi-confined, and unconfined), water quality, and geological factors (topography, drainage, formation type, rocks) all influence the requirements for implementing a certain MAR technique [59, 60]. Water from different sources (such as treated water, surface water, desalinated saltwater, precipitation, and groundwater from other aquifers) of diverse grades can be used for MAR as long as it is adequately treated before recharge.

MENA countries currently have over 2015 dams, totaling 194,800 MCM of design capacity. The climate, geology, hydrogeology, and sediment load of the dam site are some of the factors that affect how effective these dams are at recharging groundwater [61]. Sediment deposition in reservoirs can have negative effects on a number of things, including decreased dam storage capacity and recharge efficiency. MENA countries vary in the number of dams and total volume used in MAR due to factors including as location, climate, population, and economic status [55]. Saudi Arabia has the most dams (563), while Egypt has the fewest. Depending on factors such as total groundwater abstraction, climate, hydrogeological environment, and water infrastructure, groundwater renewability in the MENA area varies from less than 6–100% in a few countries [52].

In order to increase the groundwater renewability ratio, MAR has undergone significant development over the past 50 years. These developments include the use of treated sewage water as a water source for aquifer recharge, as well as the construction

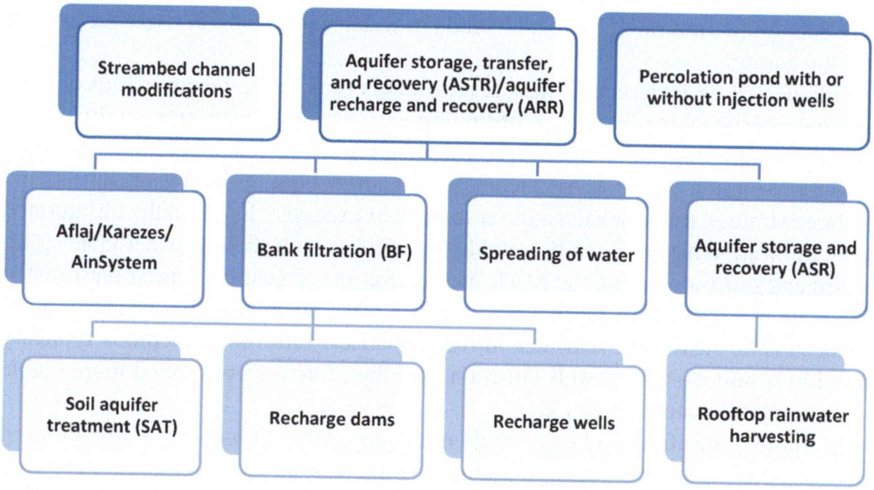

Fig. 8 Methods for managing aquifer recharge in the MENA region. Modified after [52]

of drainage wells and recharge systems for flood mitigation. The MENA regions are using planned recharging, also known as artificial recharge, more frequently as a strategy for water management [61]. Figure 8 shows the methods for managing aquifer recharge in the MENA region.

Artificial recharge to groundwater via recharge wells is a potential alternative for RWH and exploitation for enhancement of groundwater resources (increase in groundwater table), which also saves electricity. Furthermore, the history of expanding worldwide with the RWH approach is currently receiving greater focus [62]. Groundwater recharge is affected by several geological and environment factors, including soil infiltration capacity, porosity and permeability of waterbearing geological formations, the spatiotemporal distribution of rainfall, temperature, and wind conditions in the area, and vegetation types. Recharge rates vary greatly in space and time. The most essential component influencing groundwater recharge is precipitation volume [63].

Ponds are natural or intentionally constructed bodies of surface water. Rainfall in the pond infiltrates into the aquifer, raising the quality and level of groundwater. The main negative aspects of the pond as a recharge system are excessive evaporation and siltation [64]. The precipitation collected in the pond infiltrates the aquifer, enhancing groundwater level and quality. According to the researches on the pond as a recharge structure, percolation ponds should be built on permeable terrain with injection wells to reap the most benefits from this construction [65]. On the other side, the impact of the percolation pond on groundwater was assessed by Raicy and Elango [66] to compare the periodical groundwater level with rainfall. The dilution of ionic concentration in water was three times higher after the construction of recharge shaft. The long-term groundwater quality in the surrounding area of the pond improves gradually with time. The total dissolved solids (TDS) decrease considerably with time due to the dilution of dissolved solids in water with the fresh water recharging into the aquifer.

Dillon et al. [67] described how several kinds of reactive layers rich in organic material were used at the base of infiltration basins to improve water quality as soil passed through. Infiltration through constructed basins, recharge boreholes, or wells were the common viable MAR techniques. Aquifer type, groundwater levels, and recharge strategy used for each aquifer to prevent excessive losses in the unsaturated zone. Infiltration basins are only suitable for aquifers with shallow water tables [68]. In arid and semi-arid areas, the MAR method has the largest influence on estimated cost, with the order of increasing cost for 1 Mm^3/y schemes as follows: infiltration basin, ASTR, ASR, and seawater intrusion barrier. Infiltration methods typically have lower unit costs than well injection, and therefore are considered more viable for agricultural water supply [52].

In the face of climate change challenges, the global community must develop ways to become more robust to the effects. One such method that might help with resilience is ASR [69]. ASR is a water management approach that includes storing water in a suitable aquifer via a well when it is available and recovering the water from the same well when it is needed. This can help to alleviate the pressures of water shortage while also providing additional advantages such as minimizing seawater intrusion and land subsidence [60, 70]. The implementation of ASR can help to build community resilience and alleviate the impacts of drought, water scarcity, since the storage water will help to manage existing water resources to increase preparedness [71].

Soil Aquifer Treatment (SAT) is a MAR approach that allows water to penetrate from a pond. The microorganisms in the water are significantly eliminated during the transit through an unsaturated aquifer media. Changes in geochemistry and the quality of recovered water depend upon the types of materials and distance of the pumping well from the pond [72, 73]. It is estimated that a total of 16,790 m^3 of tertiary treated wastewater was released to the basins of Sulaibiya, Kuwait and only about 8000 m^3 out of 16,790 m^3 has reached the groundwater table [59]. An additional goal of soil aquifer treatment systems that periodically infiltrate recycled water is to enhance water quality.

4 Benefits of Rainwater Harvesting for MAR and Flood Mitigation

The water cycle is an important part of ecosystem functioning. Water conservation (WC) is the process of intercepting precipitation, storing it, and making it accessible for use or function in the environmental system for a certain duration of time. As an essential water ecosystem function, demand for WC is growing in the face of high social and economic demand, fast population increase and urbanization, and global climate change [74]. Spatially, WC is the regulation of the water cycle in terrestrial ecosystems at the vertical level, i.e., the interaction of precipitation and evapotranspiration with the vegetation canopy, the understory litter layer and the soil layer, which ultimately recharges the regional water supply in the form of surface runoff, subsurface runoff and loamy midstream flow or is stored within the system [75]. In the long run, this mostly takes the form of precipitation storage and flood mitigation, as well as groundwater recharge to rivers during low precipitation seasons, which may successfully safeguard water supplies, minimize floods, and sustain healthy regional ecosystems [76].

Plans for conserving water can also be put into action through improved management, planning, technology, regulation, and enforcement of water delivery systems, as well as by rehabilitating people's habits of using water. Monitoring surface water resources, ground water levels, and stream flow sources and reservoirs are a crucial component of water conservation measures [77].

Throughout thousands of years, humans have developed techniques for managing water resources in an attempt to lessen the effects of flood catastrophes on people and their belongings. Through non-engineering techniques to keep people away from floodwaters and engineering measures to keep floodwaters away from humans, societies have consistently improved flood control standards [78, 79].

The risk assessment of natural disasters includes qualitative, semi quantitative, and quantitative approaches [80]. The result of qualitative assessment is the relative magnitude of natural disaster risk. The result of semi-quantitative risk evaluation can be expressed as the multiplication of the frequency grade and consequence grade. Quantitative assessment converts the loss result into a monetary value, to obtain an expected loss, such as the expected annual loss or the cumulative loss, in order to accurately measure the impact of flood [81].

Combining diverse methods, the simulation-based methodology is able to assess damage losses, simulate the course of floods, and determine flood risk. Research on flood risk and flood losses is now being conducted using this methodology in an increasing number of studies [82]. The two components of simulation-based flood risk assessment may be separated by the risk assessment framework. A flood inundation map is produced using a hydrological and hydrodynamic model that is based on a hazard analysis in the first section. A damage assessment model based on exposure and vulnerability analysis is used in the second section to calculate disaster loss outcomes [83].

Rainwater harvesting is a low-cost approach for increasing water security and is seen as a low-cost adaptation measure to climate change. It also promises to enhance crop productivity by expanding rainwater gathering in rural regions [84]. Other advantages include reduced soil erosion, lower runoff peak flow, flood prevention, and higher groundwater recharge, among others. There are several rainwater collecting methods available that may be tailored to local climatic, biophysical, and socioeconomic circumstances and are suited for either urban or rural settings [85]. By using rainwater gathered behind recharge dams, available rainfall collection devices may boost groundwater recharge by 4–40%. Over the last sixty years, storm water and harvested rainfall have been effectively employed for groundwater recharge. Except for Oman, groundwater abstraction exceeds renewability in all GCC nations. With the current major change in precipitation patterns and intensity, MAR approaches might play a significant role in mitigating the effects of groundwater depletion [52].

The recharge from rainfall is around 2.397 BCM per year, accounting for 88% of the inflow in Oman. Dam storage capacity in Oman has risen from 87.6 MCM in 2007 to 323 MCM in 2018. Oman has 43 recharge dams with a combined storage capacity of 93.5 MCM/year as of 2021 [86, 87]. Similarly, the UAE now has over 120 recharge and storage dams, intercepting an estimated 150 MCM/year of surface water runoff from 15 major catchment basins [88, 89]. There are 313 shallow recharge wells in Doha, replenishment of urban storm water through it also known as passive recharge wells, with water entering the aquifer under the effect of gravity in shallow depressions dug by Qatar's Ministry of Environment. The majority of these recharge wells replenish 10.7 MCM/year of water, accounting for approximately 14.26% of total natural recharge [52].

MAR can provide greater benefits such as recovering a water resource that would otherwise be lost to the sea and having less environmental consequences than new surface water dams, extensive transfer pipelines, or desalination facilities. Furthermore, because recharge is primarily gravity-based and treatment requirements are confined to pre settlement, energy costs for MAR can be comparable to those for traditional groundwater abstraction [53, 90].

Ríos et al. [91] concluded the advantages of an MAR as: (1) it serves to improve the quality of reclaimed water, contributing to better resource management; (2) it functions as a permanent storage system without losses due to evapotranspiration and without the use of surface infrastructures to store water; (3) it reduces the overexploitation of aquifers by guaranteeing a higher recharge rate than the natural one, so that groundwater extraction becomes a more sustainable process; and (4) it allows combating sources contamination through marine intrusion by creating water barriers that progressively improve their quality and help to protect the resources of the aquifer itself.

MAR can be accomplished with less cost, less energy used, and the minimum negative environmental effects of building new dams or desalination facilities. By comparison with the direct reuse of treated wastewater and decreased groundwater abstraction, MAR can sustain public irrigation perimeters and offer a strategic groundwater resource for usage during periods of drought [46].

Although its effectiveness, MAR is still not a very common strategy for managing groundwater in this area. This might be because the MENA region has seen several droughts over the past three decades, which has prevented an adequate quantity of rainfall to support this kind of effort [52]. Therefore, a workable way to improve the hydrodynamic and physicochemical characteristics of the groundwater is to replenish water-table aquifers directly with precipitation collected behind dams.

5 Case Studies and Site Suitability Identification

Schild et al. [85] investigated the economic feasibility of the implementation of the two most commonly practiced rainwater harvesting applications in a rural and urban setting in the West Bank of Palestine. Eyebrow terracing in olive cultivation and domestic rooftop harvesting in urban areas were investigated. They found that rooftop harvesting can harvest, on average, about 30% of the annual domestic water demand but the construction costs are generally too high to be cost-effective for households to implement.

Ertop et al. [92] calculated to which extent the rain harvest that can be obtained from plastic greenhouses in Antalya Province (Turkey) and the results showed that rainwater harvesting can be utilized in tomato cultivation to ensure the sustainability of existing water resources and save water.

Piemontese et al. [93] evaluated 1046 case studies pertaining to rainwater harvesting, the findings indicate that around 19% of the world's farmland is capable of reproducing the successful case studies' increased crop output (i.e., demonstrating an increase in crop productivity upon implementation).

Sherif et al. [52] investigated the primary factors impacting groundwater recharge augmentation via recharge dams and MAR in MENA nations, using examples from the Kingdom of Saudi Arabia and the United Arab Emirates. They confirmed that rainwater harvesting techniques are a traditional practice in the MENA region, with only around 25% of the stored water in dam ponding sites reaching the target aquifers. This is because of reservoir sedimentation.

Alataway and El Alfy [94] studied rainwater harvesting in three small dam reservoirs and one moderate-capacity dam reservoir in the Al-Alb basin, central Saudi Arabia, to address water scarcity in arid areas. According to the study, these small dams can capture significant rainfall (50–100 mm). Smaller upstream dams reduce peak runoff by 14%, whereas the main dam increases runoff volume by 3%. Constructing recharge wells in dam reservoirs increases groundwater recharge by 44% while reducing evaporation losses by 86%. However, well efficiency decreases with time due to siltation, necessitating periodic maintenance. This study underlines the prospect of long-term water management in dry areas by rainwater harvesting and groundwater recharge.

Pachpute et al. [95] studied the sustainability of various RWH systems in Tanzania's Makanya catchment. They assessed storage-based RWH systems, like microdams, ponds, and tanks, in different conditions. Population growth upstream required RWH for water, affecting downstream water availability due to increased hillside farming. Sustainability of RWH systems was influenced by factors like rainfall, runoff, local capacity, labor, and institutional support.

Ibrahim [96] investigated novel rainwater harvesting techniques to provide additional drinking water for Gadarif city in central Sudan, addressing urban growth challenges. The results suggest that this approach can offer a valuable solution for water-scarce regions, especially semi-arid areas facing climate change and population growth, ensuring a sustainable source of safe drinking water.

Ammar et al. [97] examined the identification of suitable sites for RWH in arid and semi-arid regions. They found that crucial criteria included rainfall, slope, soil type, distance to settlements/streams, land use, and cost considerations.

Ezzeldin et al. [98] used a comprehensive approach, including GIS, remote sensing, and hydrological modeling, to pinpoint 12 optimal sites for check dams and 14 suitable locations for percolation tanks in Wadi Watir, Sinai Peninsula, Egypt.

Alrawi et al. [99] employed a multi-criteria decision analysis approach, integrating the analytic hierarchy process, GIS, and RS, to assess areas suitable for RWH in the Al-Qalamoun Basin, Syria. They found that 18.1% of the study area is suitable for rainwater harvesting (RWH) projects. Their study provides a valuable tool for decision-makers to identify optimal RWH sites and avoid unsuitable locations for dam construction.

Al-Kakey et al. [100] used GIS and AHP approaches to identify suitable sites and structures for artificial groundwater recharge in Iraq's Erbil Basin. The study's findings indicated two possible dam locations on Erbil's eastern fringes for artificial recharge structures for rainfall harvesting.

Baptista et al. [101] explored rooftop water harvesting for aquifer recharge and flood mitigation in João Pessoa, northeast Brazil, revealing annual estimates of aquifer recharge between 57 and 255 m^3/yr from 2004 to 2019. Their research highlights the potential of MAR to align stormwater management with water supply goals.

6 Challenges and Proposed Solutions

The successful implementation of rainwater harvesting and managed groundwater recharge is strongly reliant on competent engineering standards. Technical concerns include selecting appropriate recharge methods depending on geological conditions, optimizing the design of infiltration structures, and addressing potential water quality issues. In densely populated areas, the availability of suitable locations for recharge structures may be limited, necessitating novel engineering solutions to integrate these systems into existing urban landscapes. The challenges and solutions for implementing rainwater harvesting and using it for MAR are as follows.

Rainfall Pattern Variability: Irregular and unpredictable rainfall can reduce the effectiveness of rainwater gathering, resulting in changes in water availability for recharge. Rainwater harvesting systems rely primarily on constant and predictable rainfall. The difficulty derives from the variability in these patterns, which causes changes in water availability for recharge. To solve this, adaptive management systems that take into account changing climates and employ technologies capable of efficiently gathering and storing rainwater even during unpredictable rainfall periods are required.

Extreme Weather Events: Increasing frequency and intensity of extreme weather events, such as heavy storms, can overwhelm infrastructure designed for MAR, posing flood risks. The rapid increase in extreme weather events, such as heavy storms, poses a substantial challenge to MAR systems and raises the risk of flooding. To address this issue, it is essential to invest in resilient infrastructure capable of handling increasing water volumes during heavy rainfall. Early warning systems can also play an important role in delivering timely alarms, allowing communities to implement flood mitigation measures and mitigate possible damages.

Technical and engineering issues: Technical complexities include selecting suitable recharge methods based on geological conditions, optimizing the design of infiltration structures, and addressing water quality concerns. The technical complexities of rainwater gathering and regulated groundwater recharging necessitate novel engineering solutions. Developing improved filtration systems to solve water quality concerns, optimizing the design of infiltration structures based on geological conditions, and integrating technology that improves the effectiveness of recharge methods could be among the solutions.

Urbanization Constraints: In urban areas, limited available space for recharge structures can hinder implementation, requiring innovative engineering solutions to integrate these systems into existing landscapes. Integrating recharge structures smoothly into the existing landscape provides a problem in urban areas, where space is typically at a premium. Vertical or rooftop rainwater harvesting systems, green infrastructure, and underground storage are examples of innovative technical solutions that can help overcome geographical limits and make these technologies more suitable with urban surroundings.

7 Conclusions

Rainwater Harvesting (RWH) is a valuable method for bolstering water resources through Managed Aquifer Recharge (MAR) and flood prevention. RWH encompasses the collection and storage of rainwater, typically gathered from rooftops or other catchment areas. This collected rainwater can be directed towards MAR, which involves replenishing underground aquifers and improving groundwater supplies. Simultaneously, RWH serves as a proactive approach to flood prevention by capturing surplus rainfall, reducing surface runoff, and mitigating the risk of urban and rural flooding. This dual-purpose strategy not only contributes to the sustainability of water resources but also enhances climate resilience in regions grappling with water scarcity and an increased occurrence of extreme weather events. Within this chapter, we have explored various rainwater harvesting techniques, both traditional and modern, assessed their suitability in various contexts, and examined their role in mitigating flood risks. Additionally, we have highlighted the advantages of Rainwater Harvesting for MAR and Flood Mitigation, presented case studies, discussed site suitability evaluations, and addressed the challenges associated with Rainwater Harvesting, along with potential solutions.

The results of several case studies reveal that approximately 19% of the global agricultural land has the capacity to replicate successful rainwater harvesting practices, leading to improved crop productivity.

One of the primary hindrances to incorporating water harvesting into various practices in the MENA region encompasses several factors: limited access to comprehensive hydrological data, variability in rainfall patterns, constraints linked to urbanization, technical and engineering obstacles, and the consequences of extreme weather events. This scarcity poses a barrier to confidently planning, designing, and implementing water harvesting projects. Furthermore, the sustainability of RWH systems is influenced by factors such as rainfall, runoff, local capacity, labor, and institutional support. In conclusion, the advantages of the RWH technique can be summarized as follows:

- RWH is often a cost-effective alternative compared to large-scale infrastructure projects for water supply and flood control. It can be applied across various scales, from individual households to entire communities.
- RWH enables the collection and storage of rainwater, providing a dependable and sustainable source of freshwater. This reduces reliance on unsustainable groundwater extraction, supporting the long-term management of groundwater resources.
- RWH systems can be designed to directly recharge aquifers with rainwater, aiding in the restoration of groundwater levels, mitigating over-extraction, and promoting a more balanced water table.
- RWH systems can efficiently manage excess rainwater during heavy rainfall events, reducing surface runoff and the risk of flooding in urban and rural areas. This safeguards communities and infrastructure from flood-related damages.
- RWH enhances climate resilience by capturing and storing rainwater during periods of intense rainfall and droughts, guaranteeing a more stable water supply in regions susceptible to climate change.

8 Recommendations

Future studies should focus research and development projects aiming at improving the effectiveness of rainwater harvesting for managed groundwater recharge technology. This includes investigating innovative engineering solutions, sustainable materials, and improved filtering systems to address water quality challenges. Developing improved filtration systems to solve water quality concerns, optimizing the design of infiltration structures based on geological conditions, and integrating technology that improves the effectiveness of recharge methods could be among the solutions. Establishing advanced warning systems is vital in issuing prompt alerts, thus empowering communities to initiate flood control actions and lessen potential damages. Policymakers and decision-makers should work on incorporating climate-resilient infrastructure into urban planning, building systems capable of adjusting to shifting rainfall patterns and extreme weather events for more robust flood prevention. By following these recommendations, stakeholders can help to improve resilient and sustainable water management practices with effective flood mitigation measures via RWH technique.

References

1. deMenocal PB (2001) Cultural responses to climate change during the late Holocene. Science 292(5517):667–673. https://doi.org/10.1126/science.1059287
2. Zhang DD, Brecke P, Lee HF, He Y-Q, Zhang J (2007) Global climate change, war, and population decline in recent human history. Proc Natl Acad Sci 104(49):19214–19219. https://doi.org/10.1073/pnas.0703073104
3. Taylor D, Brokensha DW, Castro AP (2012) Climate change and threatened communities: vulnerability, capacity, and action, 216 p. ISBN:978-1-85339-735-6
4. Fletcher R, Penny D, Evans D, Pottier C, Barbetti M, Kummu M, Lustig T (2008) The water management network of Angkor, Cambodia. Antiquity 82(317):658–670. https://doi.org/10.1017/S0003598X00097295
5. Scarborough VL, Lucero LJ (2010) The non-hierarchical development of complexity in the semitropics: water and cooperation. Water History 2(2):185–205. https://doi.org/10.1007/s12685-010-0026-z
6. Wanner H, Beer J, Bütikofer J, Crowley TJ, Cubasch U, Flückiger J, Goosse H, Grosjean M, Joos F, Kaplan JO, Küttel M, Müller SA, Prentice IC, Solomina O, Stocker TF, Tarasov P, Wagner M, Widmann M (2008) Mid- to late Holocene climate change: an overview. Quat Sci Rev 27(19–20):1791–1828. https://doi.org/10.1016/j.quascirev.2008.06.013
7. Yannopoulos S, Giannopoulou I, Kaiafa-Saropoulou M (2019) Investigation of the current situation and prospects for the development of rainwater harvesting as a tool to confront water scarcity worldwide. Water 11(10):2168. https://doi.org/10.3390/w11102168
8. Prinz D (1996) Water harvesting—Past and future. In: Sustainability of irrigated agriculture. Springer Netherlands, pp 137–168. https://doi.org/10.1007/978-94-015-8700-6_10
9. Campisano A, Butler D, Ward S, Burns MJ, Friedler E, DeBusk K, Fisher-Jeffes LN, Ghisi E, Rahman A, Furumai H, Han M (2017) Urban rainwater harvesting systems: research, implementation and future perspectives. Water Res 115:195–209. https://doi.org/10.1016/j.watres.2017.02.056
10. Jebari S, Berndtsson R, Bahri A (2015) Challenges of traditional rainwater harvesting systems in Tunisia. Middle East Critique 24(3):289–306. https://doi.org/10.1080/19436149.2015.1046707
11. Nasr M (1999) Assessing desertification and water harvesting in the middle East and North Africa: policy implications. ZEF-Discussion Papers on Development Policy 10:59. http://www.zef.de
12. Boers ThM, Ben-Asher J (1982) A review of rainwater harvesting. Agric Water Manag 5(2):145–158. https://doi.org/10.1016/0378-3774(82)90003-8
13. Tamagnone P, Comino E, Rosso M (2020) Rainwater harvesting techniques as an adaptation strategy for flood mitigation. J Hydrol 586:124880. https://doi.org/10.1016/j.jhydrol.2020.124880
14. Angelakis AN, Mays LW, Koutsoyiannis D, Mamassis N (2012) Evolution of water supply through the Millennia. IWA Publishing. http://library.oapen.org/handle/20.500.12657/25813
15. Foltz RC (2002) Iran's water crisis: Cultural, political, and ethical dimensions. J Agric Environ Ethics 15(4):357–380. https://doi.org/10.1023/A:1021268621490
16. Reza Balali M, Keulartz J, Korthals M (2009) Reflexive water management in arid regions: the case of Iran. Environ Values 18(1):91–112. https://doi.org/10.3197/096327109X404807
17. Wessels J (2008) To cooperate or not to cooperate? Collective action for rehabilitation of traditional water tunnel systems (Qanats) in Syria. Amsterdam University Press. https://hdl.handle.net/11245/1.288733
18. Noaman A, AL-Washali T, Center E (2019) The Yemen experience on water harvesting. In: Nairobi knowledge and experience exchange symposium. www.adslgate.com
19. VS J, Jacob RS (2023) Traditional rainwater harvesting systems in the world, pp 15–33. https://doi.org/10.1007/978-3-031-38028-0_2

20. Anchan SS, Prasad HS (2021) Feasibility of roof top rainwater harvesting potential-a case study of South Indian University. Cleaner Engineering and Technology 4:100206. https://doi.org/10.1016/j.clet.2021.100206
21. Dao AD, Han M, Nguyen VA, Ho XQ, Kim TH (2009) Flooding mitigation plan at downtown of Hanoi by rainwater harvesting. In: Proceeding of UNESCO-IHP session: combining rain culture with rainwater harvesting technology for sustainable development in South and East Asia
22. Hardie M (2010) Rainwater storage gutters for houses. Sustainability 2(1):266–279. https://doi.org/10.3390/su2010266
23. Melville-Shreeve P, Ward S, Butler D (2016) Rainwater harvesting typologies for UK houses: a multi criteria analysis of system configurations. Water 8(4):129. https://doi.org/10.3390/w8040129
24. Brodie IM (2008) Hydrological analysis of single and dual storage systems for stormwater harvesting. Water Sci Technol 58(5):1039–1046. https://doi.org/10.2166/wst.2008.469
25. Gee KD, Hunt WF (2016) Enhancing stormwater management benefits of rainwater harvesting via innovative technologies. J Environ Eng 142(8). https://doi.org/10.1061/(ASCE)EE.1943-7870.0001108
26. Sample DJ, Liu J (2014) Optimizing rainwater harvesting systems for the dual purposes of water supply and runoff capture. J Clean Prod 75:174–194. https://doi.org/10.1016/j.jclepro.2014.03.075
27. Atik D (2019) Rainwater harvesting in urban and rural areas. https://www.researchgate.net/publication/338741398
28. Faza K, Suwartha N (2021) The effect of roof surface area on the quality and quantity of rainwater runoff in the rainwater harvesting system. In: IOP conference series: earth and environmental science 623(1):012010 (IOP Publishing). https://doi.org/10.1088/1755-1315/623/1/012010
29. Wahba M, Mahmoud H, Elsadek WM, Kanae S, Hassan HS (2022) Alleviation approach for flash flood risk reduction in urban dwellings: a case study of fifth District, Egypt. Urban Climate 42. https://doi.org/10.1016/j.uclim.2022.101130
30. Oweis TY (2022) Micro-catchment rainwater harvesting. In Unconventional water resources. Springer International Publishing, pp 73–94. https://doi.org/10.1007/978-3-030-90146-2_4
31. Freni G, Liuzzo L (2019) Effectiveness of rainwater harvesting systems for flood reduction in residential urban areas. Water 11(7):1389. https://doi.org/10.3390/w11071389
32. Hu M, Zhang X, Li Y, Yang H, Tanaka K (2019) Flood mitigation performance of low impact development technologies under different storms for retrofitting an urbanized area. J Clean Prod 222:373–380. https://doi.org/10.1016/j.jclepro.2019.03.044
33. Davydov R, Antonov V, Molodtsov D, Cheremisin A, Korablev V (2018) The simulation model for a flood management by flood control facilities. MATEC Web of Conferences 245:15002. https://doi.org/10.1051/matecconf/201824515002
34. Duc Tran D, van Halsema G, Hellegers PJGJ, Phi Hoang L, Quang Tran T, Kummu M, Ludwig F (2018) Assessing impacts of dike construction on the flood dynamics of the Mekong Delta. Hydrol Earth Syst Sci 22(3):1875–1896. https://doi.org/10.5194/hess-22-1875-2018
35. Karamouz M, Heydari Z (2020) Conceptual design framework for coastal flood best management practices. J Water Resour Plann Manag 146(6). https://doi.org/10.1061/(ASCE)WR.1943-5452.0001224
36. Morsy KM, Morsy AM, Hassan AE (2018) Groundwater sustainability: opportunity out of threat. Groundw Sustain Dev 7:277–285. https://doi.org/10.1016/j.gsd.2018.06.010
37. Zhao W, Ma J (2019) Landscape applications of rainwater resources in urban green space: a case study of green space reconstruction Along Riverside Street in Suzhou. IOP conference series: earth and environmental science 349(1):012020. https://doi.org/10.1088/1755-1315/349/1/012020
38. Wahba M, Hassan HS, Elsadek WM, Kanae S, Sharaan M (2023) Novel utilization of simulated runoff as causative parameter to predict the hazard of flash floods. Environ Earth Sci 82(13):333. https://doi.org/10.1007/s12665-023-11007-w

39. Wahba M, Sharaan M, Elsadek WM, Kanae S, Hassan HS (2023) Categorization of urban basin according to the runoff depth: case study of Katsushika Ward and Edogawa City Basin, Japan, pp 131–142. https://doi.org/10.1007/978-981-99-4101-8_10
40. El-Rawy M, Elsadek WM, De Smedt F (2022) Flash flood susceptibility mapping in Sinai, Egypt using hydromorphic data, principal component analysis and logistic regression. Water 14(15):2434. https://doi.org/10.3390/w14152434
41. Elsadek WM, Wahba M, Al-Arifi N, Kanae S, El-Rawy M (2023) Scrutinizing the performance of GIS-based analytical hierarchical process approach and frequency ratio model in flood prediction—Case study of Kakegawa, Japan. Ain Shams Eng J 102453. https://doi.org/10.1016/j.asej.2023.102453
42. Mansour MM, Ibrahim MG, Fujii M, Nasr M (2022) Recent applications of flash flood hazard assessment techniques: case studies from Egypt and Saudi Arabia. Adv Eng Forum 47:101–110. https://doi.org/10.4028/p-03z404
43. Mansour MM, Nasr M, Fujii M, Yoshimura C, Ibrahim MG (2023) Quantification of flash flood runoff volume using morphometric parameters towards sustainability, pp 79–91. https://doi.org/10.1007/978-981-99-4101-8_6
44. Wahba M, El-Rawy M, Al-Arifi N, Mansour MM (2023) A novel estimation of the composite hazard of landslides and flash floods utilizing an artificial intelligence approach. Water 15(23):4138. https://doi.org/10.3390/w15234138
45. Wahba M, Sharaan M, Elsadek WM, Hassan HS (2023) Building information modeling integrated with environmental flood hazard to assess the building vulnerability to flash floods. Stoch Environ Res Risk Assess. Accepted
46. Standen K, Costa L, Hugman R, Monteiro JP (2023) Integration of managed aquifer recharge into the water supply system in the Algarve Region, Portugal. Water 15:2286. https://doi.org/10.3390/w15122286
47. INOWAS (Innovative Groundwater Solutions) (2018) Recharge dams. https://www.inowas.com/recharge-dams/. Accessed 15 Oct 2023
48. Ward PJ, de Ruiter MC, Mård J, Schröter K, Van Loon A, Veldkamp T, von Uexkull N, Wanders N, AghaKouchak A, Arnbjerg-Nielsen K, Capewell L (2020) The need to integrate flood and drought disaster risk reduction strategies. Water Secur 11:100070. https://doi.org/10.1016/j.wasec.2020.100070
49. Abdalla OA, Al-Rawahi AS (2013) Groundwater recharge dams in arid areas as tools for aquifer replenishment and mitigating seawater intrusion: example of AlKhod, Oman. Environ Earth Sci 69:1951–1962. https://doi.org/10.1007/s12665-012-2028-x
50. Al-Maktoumi A, El-Rawy M, Zekri S (2016) Management options for a multipurpose coastal aquifer in Oman. Arab J Geosci 9:1–4. https://doi.org/10.1007/s12517-016-2661-x
51. Bharti DDDS (2023) Global consequences of ground water scarcity and replenishment prospects at the end of current Millennium By 3000 AD. Remittances Rev 8
52. Sherif M, Sefelnasr A, Al Rashed M et al (2023) A review of managed aquifer recharge potential in the Middle East and North Africa Region with examples from the Kingdom of Saudi Arabia and the United Arab Emirates. Water 15:742. https://doi.org/10.3390/w15040742
53. Dillon P, Stuyfzand P, Grischek T et al (2019) Sixty years of global progress in managed aquifer recharge. Hydrogeol J 27:1–30. https://doi.org/10.1007/s10040-018-1841-z
54. Bresinsky L, Kordilla J, Hector T, Engelhardt I, Livshitz Y, Sauter M (2023) Managing climate change impacts on the Western Mountain Aquifer: implications for Mediterranean karst groundwater resources. J Hydrol X 20:100153. https://doi.org/10.1016/j.hydroa.2023.100153
55. Tawfik M, Badawy A, Ali M, Amdar N, Elmahdi A (2020) Urban flash floods mitigations: enhancing the link between water, sanitation and energy services in urban settings in the MENA region. Adv Civil Eng Technol 4. https://doi.org/10.31031/ACET.2020.04.000582
56. El-Rawy M, Al-Maktoumi A, Zekri S, Abdalla O, Al-Abri R (2019) Hydrological and economic feasibility of mitigating a stressed coastal aquifer using managed aquifer recharge: a case study of Jamma Aquifer, Oman. J Arid Land 11:148–159. https://doi.org/10.1007/s40333-019-0093-7

57. Halytsia O, Vrachioli M, Janik K et al (2022) Assessing economic feasibility of managed aquifer recharge schemes: evidence from cost-benefit analysis in Poland. Water Resour Manag 36:5241–5258. https://doi.org/10.1007/s11269-022-03303-0
58. El-Rawy M, Zlotnik VA, Al-Raggad M, Al-Maktoumi A, Kacimov A, Abdalla O (2016) Conjunctive use of groundwater and surface water resources with aquifer recharge by treated wastewater: evaluation of management scenarios in the Zarqa River Basin, Jordan. Environ Earth Sci 75:1146. https://doi.org/10.1007/s12665-016-5946-1
59. Parimalarenganayaki S (2021) Managed aquifer recharge in the Gulf countries: a review and selection criteria. Arab J Sci Eng 46:1–15. https://doi.org/10.1007/s13369-020-05060-x
60. Al-Maktoumi A, Zekri S, El-Rawy M, Abdalla O, Al-Abri R, Triki C, Bazargan-Lari MR (2020) Aquifer storage and recovery, and managed aquifer recharge of reclaimed water for management of coastal aquifers. Desalin Water Treat 176:67–77. https://doi.org/10.5004/dwt.2020.25499
61. Loudyi D, Kantoush SA (2020) Flood risk management in the Middle East and North Africa (MENA) region. Urban Water J 17(5):379–380. https://doi.org/10.1080/1573062X.2020.1777754
62. Hussain F, Hussain R, Wu R-S, Abbas T (2019) Rainwater harvesting potential and utilization for artificial recharge of groundwater using recharge wells. Processes 7(9):623; https://doi.org/10.3390/pr7090623
63. Khalil K, Khan Q, Mohamed M (2022) Modeling aquifer storage and recovery in the Eastern District of the United Arab Emirates using MODFLOW. Sci Rep 12:17537. https://doi.org/10.1038/s41598-022-20470-7
64. Deng C, Bailey RT (2022) Assessing the impact of artificial recharge ponds on hydrological fluxes in an irrigated stream–Aquifer system. Hydrology 9(5):91. https://doi.org/10.3390/hydrology9050091
65. Kantoush SA, Saber M, Abdel-Fattah M, Sumi T (2022) Integrated strategies for the management of Wadi Flash Floods in the Middle East and North Africa (MENA) Arid Zones: the ISFF project. In: Sumi T, Kantoush SA, Saber M (eds) Wadi flash floods. Natural disaster science and mitigation engineering: DPRI reports. Springer, Singapore. https://doi.org/10.1007/978-981-16-2904-4_1
66. Raicy MC, Elango L (2020) Percolation pond with recharge shaft as a method of managed aquifer recharge for improving the groundwater quality in the Saline Coastal Aquifer. J Earth Syst Sci 129:63. https://doi.org/10.1007/s12040-019-1333-0
67. Dillon P, Fernández Escalante E, Megdal SB, Massmann G (2020) Managed aquifer recharge for water resilience. Water 12(7):1846. https://doi.org/10.3390/w12071846
68. Sendrós A, Urruela A, Himi M et al. (2021) Characterization of a shallow coastal aquifer in the framework of a subsurface storage and soil aquifer treatment project using electrical resistivity tomography (Port de la Selva, Spain). Appl Sci 11(6):2448. https://doi.org/10.3390/app11062448
69. Agustin F, Abdullah C, Herlina D et al (2023) Aquifer geometry identification and determination of aquifer storage and recovery (ASR) in Bandung. Proceedings of the IOP conference series: earth and environmental science, IOP Publishing 012005. https://doi.org/10.1088/1755-1315/1175/1/012005
70. Zimmerman K (2023) The role of aquifer storage and recovery (ASR) in water resource management. Goal 6: clean water and sanitation 4. https://digitalcommons.coastal.edu/goal-6-clean-water/4
71. Wu P, Comte J-C, Ma Z, Li F, Chen H (2023) Scaling of groundwater flow subject to managed aquifer recharge using injection boreholes: physical experiments and upscaled numerical models. Urban Climate 51:101651. https://doi.org/10.1016/j.uclim.2023.101651
72. Turkeltaub T, Furman A, Mannheim R, Weisbrod N (2022) Continuous monitoring of a soil aquifer treatment system's physico-chemical conditions to optimize operational performance. Hydrol Earth Syst Sci 26:1565–1578. https://doi.org/10.5194/hess-26-1565-2022
73. Amin HM, Gad AA, El-Rawy M, Abdelghany UA, Sadeek RA (2021) Assessment of wastewater contaminant concentration through the vadose zone in a soil aquifer treatment system. Appl Ecol Environ Res 19(3):2385–2403. https://doi.org/10.15666/aeer/1903_23852403

74. Che X, Jiao L, Zhu X, Wu J, Li Q (2023) Spatial-temporal dynamics of water conservation in Gannan in the Upper Yellow River Basin of China. Land 12:1394. https://doi.org/10.3390/land12071394
75. Diop M, Chirinda N, Beniaich A, El Gharous M, El Mejahed K (2022) Soil and water conservation in Africa: state of play and potential role in tackling soil degradation and building soil health in agricultural lands. Sustainability 14:13425. https://doi.org/10.3390/su142013425
76. Lapworth D, Boving T, Kreamer D, Kebede S, Smedley P (2022) Groundwater quality: global threats, opportunities and realising the potential of groundwater. Sci Total Environ 811:152471. https://doi.org/10.1016/j.scitotenv.2021.152471
77. Kurunthachalam SK (2014) Water conservation and sustainability: an utmost importance. Hydrol: Curr Res 5:1. https://doi.org/10.4172/2157-7587.1000e117
78. Schroeder DW, Tsegaye S, Singleton TL, Albrecht KK (2022) GIS- and ICPR-based approach to sustainable urban drainage practices: case study of a Development Site in Florida. Water 14:1557. https://doi.org/10.3390/w14101557
79. Ross TT, Alim MA, Rahman A (2022) Community-scale rural drinking water supply systems based on harvested rainwater: a case study of Australia and Vietnam. Water 14(11):1763. https://doi.org/10.3390/w14111763
80. Qi W, Ma C, Xu H, Zhao K, Chen Z (2022) A comprehensive analysis method of spatial prioritization for urban flood management based on source tracking. Ecol Ind 135:108565. https://doi.org/10.1016/j.ecolind.2022.108565
81. Rojas O, Soto E, Rojas C, López JJ (2022) Assessment of the flood mitigation ecosystem service in a coastal wetland and potential impact of future Urban Development in Chile. Habitat Int 123:102554. https://doi.org/10.1016/j.habitatint.2022.102554
82. Han Y, Mao L, Chen X, Zhai W, Peng ZR, Mozumder P (2022) Agent-based modeling to evaluate human-environment interactions in community flood risk mitigation. Risk Anal 42:2041–2061. https://doi.org/10.1111/risa.13854
83. Wang L, Cui S, Li Y, Huang H, Manandhar B, Nitivattananon V, Fang X, Huang W (2022) A review of the flood management: from flood control to flood resilience. Heliyon 8. https://doi.org/10.1016/j.heliyon.2022.e11763
84. Rawat A, Panigrahi N, Yadav B et al. (2023) Scaling up indigenous rainwater harvesting: a preliminary assessment in Rajasthan, India. Water 15(11):2042. https://doi.org/10.3390/w15112042
85. Schild JEM, Fleskens L, Riksen M, Shadeed S (2023) Economic feasibility of rainwater harvesting applications in the West Bank, Palestine. Water 15(6):1023. https://doi.org/10.3390/w15061023
86. Prathapar SA, Bawain AA (2014) Impact of sedimentation on groundwater recharge at Sahalanowt Dam, Salalah, Oman. Water Int 39:381–393. https://doi.org/10.1080/02508060.2014.895889
87. Al-Maktoumi A (2018) Silting of recharge Dams in Oman: problems and management strategies. In: Hydrolink 2018/4. Madrid: international association for hydro-environment engineering and research (IAHR), pp 115–117
88. Sherif MM, Mohamed MM, Shetty A, Almulla M (2011) Rainfall-runoff modeling of three wadis in the northern area of UAE. J Hydrol Eng 16:10–20. https://doi.org/10.1061/(ASCE)HE.1943-5584.00002
89. Sherif M, Ebraheem A, Shetty A (2017) Groundwater recharge from Dams in United Arab Emirates. In: Proceedings of the world environmental and water resources congress, pp 139–46
90. Ebrahim GY, Lautze JF, Villholth KG (2020) Managed aquifer recharge in Africa: taking stock and looking forward. Water 12(7):1844. https://doi.org/10.3390/w12071844
91. Ríos IH, Cruz-Pérez N, Chirivella-Guerra JI et al (2023) Proposed recharge of Island aquifer by deep wells with regenerated water in Gran Canaria (Spain). Groundw Sustain Dev 22:100959. https://doi.org/10.1016/j.gsd.2023.100959
92. Ertop H, Kocięcka J, Atilgan A, Liberacki D, Niemiec M, Rolbiecki R (2023) The importance of rainwater harvesting and its usage possibilities: Antalya example (Turkey). Water 15(12):2194. https://doi.org/10.3390/w15122194

93. Piemontese L, Castelli G, Fetzer I, Barron J, Liniger H, Harari N, Bresci E, Jaramillo F (2020) Estimating the global potential of water harvesting from successful case studies. Glob Environ Chang 63:102121. https://doi.org/10.1016/j.gloenvcha.2020.102121
94. Alataway A, El Alfy M (2019) Rainwater harvesting and artificial groundwater recharge in arid areas: case study in Wadi Al-Alb, Saudi Arabia. J Water Resour Plan Manag 145(1):05018017. https://doi.org/10.1061/(ASCE)WR.1943-5452.0001009
95. Pachpute JS, Tumbo SD, Sally H et al (2009) Sustainability of rainwater harvesting systems in rural catchment of Sub-Saharan Africa. Water Resour Manage 23:2815–2839. https://doi.org/10.1007/s11269-009-9411-8
96. Ibrahim MB (2009) Rainwater harvesting for urban areas: a success story from Gadarif City in Central Sudan. Water Resour Manag 23(13):2727–2736. https://doi.org/10.1007/s11269-009-9405-6
97. Ammar A, Riksen M, Ouessar M, Ritsema C (2016) Identification of suitable sites for rainwater harvesting structures in arid and semi-arid regions: a review. Int Soil Water Conserv Res 4(2):108–120. https://doi.org/10.1016/j.iswcr.2016.03.001
98. Ezzeldin M, Konstantinovich SE, Igorevich GI (2022) Determining the suitability of rainwater harvesting for the achievement of sustainable development goals in Wadi Watir, Egypt using GIS techniques. J Environ Manag 313:114990. https://doi.org/10.1016/j.jenvman.2022.114990
99. Alrawi I, Chen J, Othman AA, Ali SS, Harash F (2023) Insights of Dam site selection for rainwater harvesting using GIS: a case study in the Al-Qalamoun Basin, Syria. Heliyon 9(9). https://doi.org/10.1016/j.heliyon.2023.e19795
100. Al-Kakey O, Othman A, Merkel B (2023) Identifying potential sites for artificial groundwater recharge using GIS and AHP techniques: a case study of Erbil Basin, Iraq. Kuwait J Sci 50(1B). https://doi.org/10.48129/kjs.11917
101. Baptista VS, Coelho VH, Bertrand GF, da Silva GB, Caicedo NO, Montenegro SM, Stefan C, Glass J, Heim R, Conrad A, Almeida CD (2023) Rooftop water harvesting for managed aquifer recharge and flood mitigation in tropical cities: towards a strategy of co-benefit evaluations in João Pessoa, Northeast Brazil. J Environ Manag 342:118034. https://doi.org/10.1016/j.jenvman.2023.118034

Soil Aquifer Treatment (SAT) for Managed Aquifer Recharge and Water Quality Improvement in the MENA Region

Hani Alharbi and Mustafa El-Rawy

Abstract Water scarcity in arid and semi-arid regions, notably in the Middle East and North Africa (MENA), poses significant challenges for both human populations and natural ecosystems. These regions face difficulties due to low rainfall, high rates of evaporation, and a precarious balance between water supply and demand, making water scarcity a critical issue impacting various development aspects. Soil Aquifer Treatment (SAT) used for Managed Aquifer Recharge (MAR) offers a sustainable approach for long-term water management. In the context of increasing global water shortages, SAT stands out as a resilient and environmentally friendly method that not only replenishes aquifers but also enhances water quality. This chapter aims to contribute to the expanding knowledge base on SAT, providing valuable insights for scholars, policymakers, and practitioners involved in sustainable water resource management. Additionally, the chapter discusses challenges and recommendations for implementing SAT technology in the MENA region. SAT could be a particularly viable solution for enhancing water resources and improving the partially treated wastewater in MENA countries, especially in developing nations.

Keywords Soil aquifer treatment · Water quality improvement · Wastewater treatment · Water management · Soil parameters

H. Alharbi
Civil Engineering Department, College of Engineering, Shaqra University, Dawadmi 11911, Saudi Arabia
e-mail: h.alharbi@su.edu.sa

M. El-Rawy (✉)
Civil Engineering Department, Faculty of Engineering, Minia University, 61111 Minia, Egypt
e-mail: mustafa.elrawy@mu.edu.eg

M. El-Rawy and A. Negm (eds.), *Managed Aquifer Recharge in MENA Countries*, Earth and Environmental Sciences Library,
https://doi.org/10.1007/978-3-031-58764-1_4

1 Introduction

The scarcity of freshwater is a global issue, with 40% of the world's population expected to be affected by water scarcity by 2030 [1]. Unfortunately, the majority of Earth's water is salty, found in oceans and seas, and only a small fraction, less than 3%, constitutes freshwater with acceptable salinity levels [2, 3]. Moreover, less than 1% of this freshwater is readily usable for essential purposes such as drinking, agriculture, and industry, since the remainder is frozen as ice [4].

The Middle East and North Africa (MENA) region confronts a critical challenge. Home to some of the most water-scarce countries globally, the MENA region grapples with a unique confluence of environmental, climatic, and hydrological conditions. Therefore, the situation is aggravated by low precipitation, high evaporation rates, and over-reliance on unsustainable groundwater sources [5, 6]. Moreover, the region, experiencing rapid population growth and climate change impacts, is nearing a critical water scarcity threshold. With a population forecast to exceed 720 million by 2050, the region faces diminishing water resources and escalating stress on freshwater availability [7]. The water scarcity is so acute that, in some areas, freshwater withdrawals are significantly higher than the available renewable water resources, primarily due to the use of fossil groundwater, desalination, and wastewater reuse [7].

Climate change is poised to exacerbate water scarcity in the MENA region. With expected increases in average global temperatures and varying precipitation levels, the hydrological cycle will be significantly affected, leading to changes in water quality and availability [8]. This situation is particularly alarming in countries like Saudi Arabia, where the combination of climate factors and water resource mismanagement intensifies the water crisis [8]. Research conducted in various regions of Saudi Arabia has revealed that a significant amount of groundwater quality exceeds acceptable standards, leading to the conclusion that most of this water is not suitable for drinking [9–11].

The demand for freshwater, driven by population growth, urbanization, and industrialization, makes finding sustainable water sources imperative. In addressing water scarcity, wastewater reclamation emerges as a critical non-traditional water resource. With a growing interest in wastewater reclamation since the mid-forties [12], countries in the MENA region like Egypt, Jordan, Oman, and Tunisia have integrated treated wastewater into their development resources, especially for agricultural irrigation [13–16]. The quality of wastewater, varying significantly based on its source and treatment level, poses challenges and opportunities for sustainable water management [8, 12, 13, 16, 18].

The importance of treated wastewater reuse (WWR) in integrated water resources management (IWRM) is increasingly recognized in policy circles. Countries in the MENA region are focusing on regulatory frameworks and national plans to maximize the use of treated water, as seen in the examples of Saudi Arabia, Egypt, and Abu Dhabi. These efforts underscore the need for comprehensive and consistent regulatory frameworks to promote sustainable water reuse practices.

In light of these challenges, this chapter delves into Soil Aquifer Treatment (SAT) as a method for managed aquifer recharge and water quality improvement. It will explore the principles, applications, and feasibility of SAT in the MENA region, along with case studies and strategies to mitigate the challenges of implementing SAT. The chapter aims to provide a comprehensive understanding of SAT's role in addressing the water scarcity crisis in the MENA region.

2 Soil Aquifer Treatment (SAT) System

The SAT system functions as a natural treatment approach for pre-treated wastewater. It encompasses the process of wastewater infiltration through recharge basins, percolation into the permeable soil within the unsaturated zone, ultimately replenishing the groundwater aquifer. Subsequently, the water is recovered through extraction wells for various purposes [19–21], as shown in Fig. 1. SAT serves as an effective method for enhancing water quality through various physical, chemical, and biological removal mechanisms [17, 18, 22].

Upon recharge into aquifers, vital processes such as filtration, biological degradation, adsorption, and ion exchange occur, leading to effluent purification before dispersion within the aquifer [16, 23, 24]. SAT is a crucial technique for managing and conserving natural groundwater resources [21, 25].

Historically, SAT has been in practice for over a century, with notable implementations such as the Montebello Forebay project in Los Angeles County, California, and the use of forested recharge basins in the Rhine River since 1920 [18, 22]. These examples highlight SAT's long-standing application in water management.

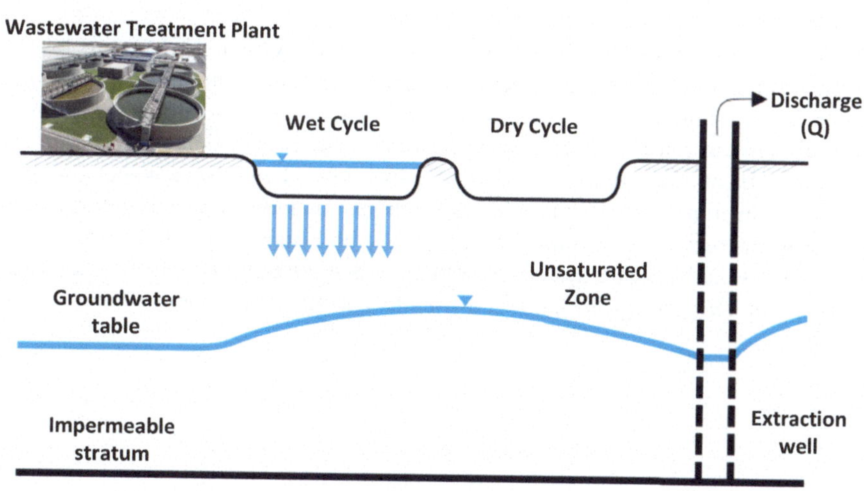

Fig. 1 Illustrates the SAT system for pre-treated wastewater. Modified after [19]

SAT is recognized as a type of Managed Aquifer Recharge (MAR) system and is particularly valuable for its ability to remove multiple contaminants [26, 27]. It has been adopted globally, not only in arid and semi-arid regions where groundwater resources are overexploited but also in developed and developing countries due to its robustness, environmental friendliness, and minimal reliance on chemicals and energy.

A key feature of SAT is its dual functionality in both treating wastewater and storing it within aquifers. The technology is characterized by the dominant treatment process in the vadose zone and storage within the saturated zone. It is known for effectively removing total nitrogen, viruses, organics, and trace organic compounds, often renovating wastewater effluent to drinking water levels. Therefore, SAT is a critical component in indirect potable reuse systems [17, 28].

The primary treatment mechanisms in SAT systems include filtration, adsorption, and biological processes such as biodegradation, which occur as the wastewater infiltrates through the unsaturated soil zone [17]. These processes result in the improvement of water quality, making SAT an innovative approach for safe wastewater reclamation and reuse, including indirect potable reuse.

The practice of reclaiming and reusing treated effluent from SAT is especially beneficial in dry, arid regions. Research conducted in places like Arizona and California [29] has shown that this reclaimed water can be safely used for irrigation on all kinds of crops without restrictions. Additionally, the option of recharging groundwater is appealing. However, careful consideration must be given to the nitrogen content in cases where the aquifers are sources of drinking water.

2.1 Components of SAT System

The figure below illustrates the components of a SAT system. The process begins with a water source, which undergoes pre-treatment to remove impurities. The water is then directed into the soil aquifer system, where it is naturally purified as it percolates through the soil layers. Following this, the water may undergo post-treatment to ensure it meets the required standards for its intended end-use, whether for agricultural, industrial, or potable use. The system also includes robust monitoring and control infrastructure to maintain water quality and manage the flow throughout the process. Finally, treated water is stored in recovery facilities, ready for use when needed, completing the cycle of this eco-friendly and efficient water treatment system (Fig. 2).

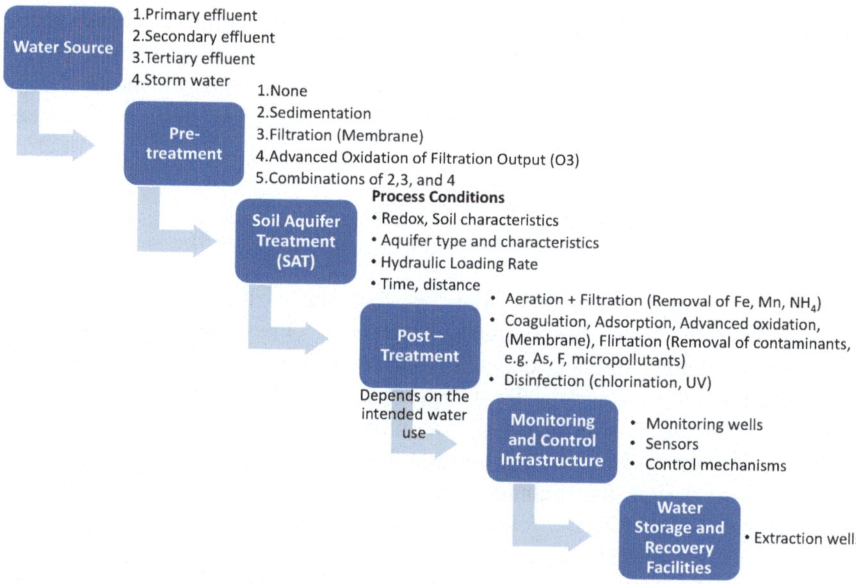

Fig. 2 Components of a SAT system Modified after Sharma and Kennedy [17]

3 Constituents of Wastewater and Their Elimination Processes

Understanding the intricate interactions between soil matrix and wastewater constituents is a crucial aspect of SAT system. These systems, designed to treat wastewater through natural filtration and biological processes in the soil, rely on a comprehensive understanding of how various wastewater elements interact with soil components. In this section, we delve into several key constituents of wastewater and their relevance to SAT system.

3.1 Biochemical Oxygen Demand (BOD)

All forms of land-based treatment methods are highly effective in eliminating biodegradable organic substances, typically measured as biochemical oxygen demand (BOD_5). The mechanisms responsible for this removal include filtration, absorption, adsorption, along with biological reduction and oxidation processes. In system, most of these processes take place either on the soil surface or in the upper layers of the soil, where there is a high level of microbial activity. The treatment processes involving oxidation-reduction reactions predominantly occur in the top third of the slope in overland flow sites. To ensure the reestablishment of aerobic conditions within the soil profile and to maintain the soil's ability to absorb water, it's essential to apply wastewater intermittently or in cycles in these systems [30].

In order to determine the appropriate quantity of degradable organic matter for land application, it's necessary to calculate the BOD loading rate. This rate is defined as follows:

$$L_{BOD} = \frac{kg\, of\, BOD\, applied\, per\, day}{(area\, loaded\, per\, day)(Cycle\, time)} \quad (1)$$

BOD_5 is measured through a 5-day test that gauges the oxygen required by microorganisms to break down organic substances. Other faster and often more reliable tests include Chemical Oxygen Demand (COD), which typically registers higher values than BOD, and the Total Organic Carbon (TOC) test. The TOC values can range from being higher than BOD in untreated wastewater to lower in treated [31] The treatment process for BOD occurs throughout various stages: the loading (or application) period, the drainage phase, and the reaeration (drying or resting) period or cycle. To ensure the soil remains in an aerobic state, the reaeration rate within each cycle should be equal to or greater than the rate of BOD application. A "rational" model, which predicts the reaeration rate based on soil conditions, depth of application, and reaeration period, has been developed [32].

Table 1 provides a summary of various instances where land treatment processes have successfully removed BOD from municipal wastewater. The data for these studies predominantly comes from long-term effect studies conducted by various researchers and organizations [33, 34].

The primary mechanism for BOD removal in these processes is biological, and this mechanism has the capability to renew itself continuously. This renewal is contingent upon the maintenance of a proper loading rate and cycle that ensures the preservation or restoration of aerobic conditions within the system.

Further insights were gained from laboratory studies in 1998 using soil columns. These studies revealed that the reduction of BOD_5 to low background levels was not affected by the level of pretreatment, the type of soil, or the rate of infiltration [30].

Table 1 BOD Removal from Municipal Wastewater via land treatment at various locations in the United States of America (USA)

		BOD_5		
Process/ location	Hydraulic loading (m/yr)	Applied (mg/L)	Soil water drainage (mg/L)	Sample depth (m)
Lake George, New York	43	38	1.2	3.2
Phoenix, Arizona	110	15	1.0	9
Hollister, California	15	220	8.0	7.6

Adapted from [35]

These findings reinforce the data presented in Table 1 and also highlight an important observation: high levels of pretreatment are not essential for the effective removal of BOD_5 in municipal land treatment systems.

3.2 *Total Suspended Solids*

Total Suspended Solids (TSS) are usually crucial development of municipal land treatment systems. SAT systems are particularly effective at removing suspended solids, primarily through soil profile filtration.

3.3 *Oil and Grease*

Oil and grease, collectively referred to as fats, oil, and grease, generally do not pose a significant concern in the land treatment of standard municipal wastewaters, unless there is an accidental spill within the municipal collection system. In such cases, designing the land treatment system specifically to handle these emergencies is not necessary, as standard containment and clean-up procedures are typically sufficient to address them. However, oil and grease are more commonly encountered as routine constituents in industrial wastewater streams. The most prevalent sources of these substances include petroleum, as well as animal and vegetable oils [30].

3.4 *pH*

The pH range that is most conducive for biological treatment typically ranges from 5 to 9, as noted by Crites and Tchobanoglous [35]. Soil possesses a considerable buffering capacity, allowing it to moderate the pH of wastewater. This capacity ensures that the efficiency of biological treatment is not compromised. The organic acids found in food processing wastewater are easily broken down, and therefore do not present a significant challenge in wastewater treatment.

Crops, similarly, can withstand a fairly wide range of pH levels. The optimal pH for their growth is reported to be between 6.4 and 8.4. When soil pH is low, there's a risk of metals becoming more soluble, which can lead to the potential leaching of these metals into groundwater. Maintaining a soil pH of 6 or higher is generally deemed sufficient to prevent the uptake of most metals by crops, as indicated by [36]. In municipal effluent, metal concentrations are typically much lower than the levels of concern. However, if there are concerns about excessive metal uptake, further precautions may be considered [36].

3.5 Pathogenic Organisms

In land treatment systems, the primary pathogens of concern include parasites, bacteria, and viruses. These pathogens can potentially contaminate groundwater, crops, and may be ingested or translocated by grazing animals. Additionally, human exposure to these pathogens can occur through offsite transmission via aerosols or runoff. The methods employed to remove pathogens in these systems are varied and include adsorption, desiccation, radiation (particularly ultraviolet light), filtration, predation, and decay due to exposure to sunlight and other unfavorable conditions.

Fecal coliform bacteria are commonly used as indicators of fecal contamination, which can originate from livestock and other warm-blooded animals. It's not unusual to detect background concentrations of fecal coliforms at levels of 100 or higher. The SAT process typically achieves a reduction of two to three orders of magnitude of fecal coliforms over several meters of travel [37].

3.6 Metals

The process of metal removal in soil is intricate, involving a combination of adsorption, precipitation, ion exchange, biogeochemical reactions, uptake by plants and microorganisms, and complexation. The adsorption of most trace elements predominantly occurs on the surfaces of clay minerals, metal oxides, and organic matter. Consequently, soils with finer textures and higher organic content tend to have a longer duration of water-soil contact, enhancing the adsorption process [30].

SAT systems are effective in metal removal, but they generally require a longer travel distance through the soil. This longer distance is necessary due to the higher hydraulic loadings and coarser soil textures associated with SAT systems [30].

3.7 Nitrogen

The process of nitrogen removal in land treatment systems is multifaceted and dynamic, largely due to the various forms of nitrogen (such as N_2, organic N, NH_3, NH_4, NO_2, NO_3) and the relative ease with which it can transition between different oxidation states. In typical municipal wastewater, nitrogen is usually found as organic nitrogen (constituting about 40%) and as ammonia/ammonium ions (approximately 60%) [30]. Activated sludge and other advanced biological treatment processes can be engineered to convert all the ammonia ion into nitrate, a process known as nitrification. However, in practice, only a portion of the ammonia nitrogen undergoes nitrification, leaving the majority of it in the effluent in the form of ammonium (with ammonia and ammonium being used interchangeably in this context) [30].

Given the health risks associated with excessive nitrogen, it's crucial in designing SAT system to accurately identify not only the total nitrogen concentration in the wastewater but also its specific forms, such as organic, ammonia, nitrate, etc. Experience with these land treatment processes has shown that the effectiveness of nitrogen retention and removal is greater when the nitrogen enters the system in a less oxidized state. This indicates that managing the oxidation state of nitrogen before it enters the land treatment system can significantly influence the efficiency of its removal [30].

4 Water Movements in Soil and Groundwater

The soil's hydraulic capacity to absorb and transmit water is a critical factor in designing SAT systems. This capacity is influenced by the soil's physical, chemical, and microbial properties, which collectively determine how water moves through it. In this section, we focus on several key hydraulic factors that are essential for SAT system. These factors include [30]:

1. Infiltration: This is the process by which water enters the soil from the surface. It is a crucial aspect as it initiates the treatment process in land-based systems.
2. Vertical Permeability (Percolation): This refers to the soil's ability to transmit water downward through its profile. Vertical permeability is important for ensuring that water can percolate through the soil layers, facilitating treatment processes as it moves.
3. Horizontal Permeability: While vertical permeability concerns the downward movement of water, horizontal permeability is about the water's ability to move sideways within the soil. This factor is vital in preventing the over-saturation of certain soil areas and ensuring an even distribution of water.
4. Groundwater Mounding: This phenomenon occurs when infiltrating water accumulates above the groundwater table, creating a mound. It's important to consider this in system design to avoid waterlogging and potential surface runoff.
5. Relationship Between Predicted Capacity and Actual Operating Rates: Understanding how the theoretical or predicted hydraulic capacity of the soil aligns with the actual operating rates observed in practice is vital. This comparison helps in assessing the efficiency and effectiveness of the system under real-world conditions.

4.1 Soil Properties

The hydraulic behavior of soil systems is influenced by a combination of their physical, biological, and chemical properties. Key physical properties include the texture, structure, and depth of the soil. On the chemical side, factors like soil pH, buffer capacity, redox potential, organic matter content, cation exchange capacity, exchangeable sodium percentage, and background nutrient levels are significant.

4.1.1 Physical Properties

Soil's physical properties are primarily concerned with its solid particles and the way these particles are aggregated. Soil texture is a term used to describe the size and distribution of these soil particles. On the other hand, soil structure refers to the manner in which these particles are aggregated or grouped together. These two aspects, texture and structure, play a crucial role in determining the soil's capacity to retain and transport water and air.

Texture

Soil texture, determined by the relative proportions of sand, silt, and clay particles, is crucial in classifying soils for various land treatment systems. The size range for these particles is clearly defined, with sand being the largest and clay the smallest. USDA classification system uses a textural triangle to classify soils based on their particle size distribution. This classification impacts soil drainage and water retention, affecting suitability for different treatment systems. Moderately coarse-textured soils are ideal for SAT systems due to their high-water acceptance and low moisture retention. However, coarse-textured soils that contain a significant amount of silt or clay (more than 10%) are not desirable for SAT systems due to potential drainage and infiltration issues [30].

Structure

Soil structure, defined by the shape and degree of aggregation of soil particles, significantly affects water and air movement, heat transfer, and soil porosity. The presence of well-defined pores and aggregates is a key indicator of good soil structure. Soils that maintain their aggregate structure when wet or tilled are considered well-structured. These soils, with their large pores, are ideal for effective water and air conduction, making them highly suitable for infiltration purposes. Generally, a well-structured soil exhibits greater permeability compared to unstructured materials of the same texture. SAT systems are particularly well-suited to soils that are sandy or loamy sand, due to these structural characteristics [30].

Soil Depth

Adequate soil depth is crucial for several reasons in land treatment systems: it enables the retention of wastewater constituents on soil particles, supports plant root development, and facilitates microbial action. In SAT systems, sufficient depth is also necessary to separate the treatment zone from saturated soil layers, as the retention of wastewater constituents is influenced by the wastewater's residence time in the soil, which in turn depends on application rate and soil permeability.

The minimum acceptable soil depth varies depending on the type of land treatment process. For SAT systems, the ideal soil depth is at least 1.5 m (5 feet), and preferably more, up to 3 m (5–10 feet) [30].

4.1.2 Chemical Properties

Chemical characteristics of soil play a vital role in influencing plant growth, the renovation of wastewater, and potentially impacting soil's hydraulic conductivity. The pH level of soil is important for both plant and bacterial growth and also affects the retention of elements like phosphorus in the soil. The redox potential (pE) of soil determines the presence of either oxidized or reduced forms of chemical elements. Organic matter in the soil can enhance its structure, which in turn can improve hydraulic conductivity. Conversely, sodium presence can adversely affect hydraulic conductivity by causing clay particles to disperse and disrupt the structural integrity necessary for water movement. Before designing a system, it's crucial to assess the soil's chemical properties to gauge its ability to support plant growth and effectively renovate wastewater [30].

Soil pH

Soil pH is often referred to as the "master variable" due to its impact on the soil's chemical, biological, and physical properties. It is affected by various factors including precipitation, irrigation water, the dissociation of carbonic acid, organic matter, mineral weathering, bio-uptake and release, aluminum hydroxy polymers, and the use of nitrogen fertilizers [38]. The pH level plays a crucial role in determining the solubility of compounds, the activity of microorganisms, and the binding of ions to exchange sites in the soil.

The availability of elements for root uptake, including both nutrients and metals, and thus crop growth, can be limited by soil pH. It also influences the activity of soil microorganisms. Chemical solubility, biochemical breakdown by microorganisms, and adsorption to soil particles are all pH-dependent processes that affect the movement of chemical constituents in the soil [30].

Soil physical properties are influenced by pH through its effect on clay dispersion and soil aggregate formation. The buffering capacity of soil is essential to prevent significant pH fluctuations that could harm plants and soil microorganisms. This buffering is mostly provided by cation exchange, or the gain or loss of H^+ ions at pH-dependent exchange sites on clay and humus particles. A well-buffered soil, containing more organic matter and/or highly charged clay, has a greater buffering capacity compared to moderately buffered soil [39].

Soil organic matter, with its numerous reactive sites, allows hydrogen ions to associate and dissociate. The surfaces of clay minerals and humus also have exchangeable ions that can interact with hydrogen ions. Thus, the cation exchange capacity (CEC) of a soil, which is the total amount of exchangeable cations it can hold, is a key factor in determining the soil's buffering capacity [40].

Soil Redox

Soil redox potential (Eh) measures the reduction and oxidation states of chemical elements in soil, significantly influencing soil aeration. It depends on the presence of oxidizing agents like oxygen and the soil's pH. Measured in volts using an electrode, Eh reflects the electron activities of chemical species in soil. It can also be expressed as pE, a dimensionless parameter related to Eh by a specific equation involving the redox potential, universal gas constant (R), temperature (T in Kelvin), and Faraday constant (F) [30].

Soil redox affects both chemical and microbial species in the soil, impacting the mobility of chemical constituents and wastewater renovation. Soil pE also indirectly influences soil structure through its effect on microbial activity. In well-aerated soils, oxidized states like Fe (III) and nitrate $\left(NO_3^-\right)$ predominate, while reduced forms like Fe (II) and ammonium $\left(NH_4^+\right)$ are found in poorly aerated soils. Lower pE values indicate reducing conditions, while higher values indicate oxidizing conditions. In the soil environment, pE values typically range just below +13.0 to near −6.0 [38].

Redox reactions in soil primarily affect elements like carbon, nitrogen, oxygen, sulfur, manganese, and iron. Oxygen can be reduced to water in soils with pE below +11.0. Below pE + 5.0, oxygen is consumed by aerobic microorganisms. In the absence of oxygen, nitrate reduction occurs below pE + 8.0, with denitrifying bacteria functioning in the pE range of +10 to 0. Iron and manganese reduction occur between pE + 7 and +5, with iron reduction happening only after oxygen and nitrate depletion, whereas manganese can reduce even in the presence of nitrate. Soils become anoxic below pE + 2.0, and sulfate reduction, driven by anaerobic microorganisms, happens when pE is less than 0. Sulfate-reducing bacteria do not grow at pE values above + 2.0 [30].

Organic Matter

Soil organic matter (SOM), typically ranging from 0.5 to 5% by weight in the surface of mineral soils, can reach up to 100% with the addition of fertilizers [39]. SOM plays a crucial role in enhancing soil structure and aggregate formation, which in turn increases soil water retention due to improved infiltration rates and water-holding capacity. It serves as an energy source for soil microorganisms, aiding in aggregate formation. Decaying organic matter, or humic substances, interact with silicate clay particles and iron and aluminum oxides to create bridges between soil particles, influencing soil pH and buffer capacity [30].

SOM has a high specific surface area, contributing significantly to the soil's cation exchange capacity (CEC). This makes it an important absorber of plant nutrients, metal cations, and organic chemicals. The presence of organic matter in soil also affects the uptake and availability of plant nutrients, especially micronutrients. Furthermore, SOM forms stable complexes with polyvalent cations like Fe^{3+}, Cu^{2+}, Ca^{2+}, Mn^{2+}, and Zn^{2+}, reducing the uptake of metals by plants and their mobility in the soil [30].

4.1.3 Water Movement in Soil

Estimating the hydraulic properties of a site is essential for the design of land treatment systems. The ability of soils to absorb and transmit water is a critical consideration, particularly for SAT. Water movement within soil is typically categorized into two types: saturated flow and unsaturated flow.

Infiltration rate

The infiltration rate, measured in millimeters per hour (mm/h) or inches per hour (in/h), indicates how quickly water enters the soil. This rate typically diminishes over time, from its initial value at the start of water application. It is influenced by the soil's texture and structure, with coarse-textured soils having higher infiltration rates due to larger pore spaces, while fine-textured or compacted soils exhibit lower rates [30].

The initial infiltration rate varies depending on the soil's moisture level [30]; drier soils initially absorb water more quickly. Factors like soil compaction, water absorption by clay particles, and the ionic composition of the soil-water also affect infiltration rates. Other influences include vegetation type, water application rate and duration, and tillage practices. Soil infiltration can be reduced due to suspended solids in wastewater, soil particle dispersion, biological growths, gas production by soil microbes, and air entrapment during wetting.

Field tests are necessary to determine steady intake rates, and these are influenced by real-world conditions in land treatment systems, often resulting in lower hydraulic loadings than those predicted by clean water infiltration rates. Good soil management can maintain or improve operating rates, while poor practices can significantly decrease them.

Soil infiltration rates can be measured in the field or estimated using mathematical models. The US EPA [30] has funded research to develop methods for quantifying soil water movement, including empirical models (like Kostiakov and Horton), mechanistic approaches (such as Green-Ampt models, Richard's equation models, and the Philips model), and evaluations of these models under different site conditions.

Permeability

Permeability, also referred to as hydraulic conductivity, is defined as the rate of water flow driven by a unit hydraulic gradient. This property is intrinsic to the soil and remains unaffected by the gradient itself, marking a key distinction between permeability and infiltration. Vertical permeability, often called percolation, is distinct from lateral flow, which depends on both the gradient and horizontal permeability (usually different from percolation rate). The primary factors influencing permeability are the physical properties of the soil. While changes in water temperature can slightly affect permeability, its impact is relatively minor compared to the influence of soil's physical characteristics [41].

5 Site Requirements

SAT is considered based on its efficacy in removing wastewater constituents, meeting discharge quality criteria, and suitability given the site characteristics. The process involves a two-phased approach [30]:

Phase 1: This initial phase is about identifying potential sites by screening available data and experience. If suitable sites for SAT are identified, the process moves to the next phase.

Phase 2: This phase entails detailed process considerations, including field investigations, preliminary design, and cost estimates. The alternatives are evaluated, and the most economical and appropriate SAT system is selected.

5.1 *Land Requirements*

For SAT systems, preliminary land requirements are estimated considering the wastewater's characteristics and climate conditions. Typical municipal wastewater characteristics necessary for the design of SAT systems are shown in Table 2. These parameters help determine the capacity and design of the SAT system. Climate factors also play a role in site selection. Included factors like water balance, growing season length, and storage capacity requirements. Climatic data must be assessed before finalizing the design criteria [31].

Site Selection

The criteria for evaluating potential sites for SAT systems, categorizing site features as preferable, moderately suitable, or unsuitable based on various geophysical and logistical factors presented below mentioned by Sharma and Kennedy [17].

Site Slope: The slope of a site is a fundamental consideration for SAT systems. A gradient of 0–5% is preferred because it facilitates uniform infiltration and minimizes runoff and erosion, thereby ensuring efficient wastewater percolation. Slopes between 5 and 15% are deemed moderately suitable; however, they may require additional engineering measures to prevent runoff. Slopes greater than 15% are generally unsuitable due to the high potential for erosion and uneven percolation of the wastewater [17].

Thickness of Vadose Zone: The vadose zone, or unsaturated zone above the aquifer, should ideally exceed 5 m in thickness to provide adequate filtration and attenuation of contaminants before the effluent reaches the groundwater. Sites with

Table 2 Typical constituent concentrations in municipal wastewater [35]

Constitutes	Concentration, g/m^3 (mg/L)
BOD	210
Suspended Solids	210
Nitrogen, total	35
Organic nitrogen	13
Ammonia nitrogen	22
Phosphorus, total	7
Potassium	15

less than 5 m of vadose zone are considered unsuitable, as the shorter filter path may not adequately protect groundwater from potential contamination [17].

Soil Type: Soil texture significantly impacts the infiltration rate and treatment efficacy of SAT systems. Sandy soils are preferred due to their high permeability, which allows for rapid infiltration and good aeration, essential for microbiological degradation processes. Loamy soils are moderately suitable, offering a balance between water retention and permeability. Clayey soils are unsuitable due to their low permeability, which can lead to surface ponding and reduced treatment effectiveness [17].

Soil Profile: Homogeneity within the vadose zone is preferred as it promotes consistent infiltration and treatment. A heterogeneous profile with less than 2 m of impermeable layers like clay or silt above the permeable layer is considered moderately suitable. The presence of clay fractions greater than 10% in the soil profile renders the site unsuitable due to the potential for reduced infiltration and increased runoff [17].

Permeability of Most Restrictive Layer: The permeability of the soil layer that presents the greatest resistance to flow should exceed 5 cm/h to be considered preferable for SAT. Rates between 1.5 and 5 cm/h are moderately suitable, although they may limit the infiltration rate and require a larger area to achieve the desired treatment. Permeability rates lower than 1.5 cm/h are unsuitable as they can hinder effective infiltration and may lead to surface ponding and hydraulic overload of the system [17].

Proximity to Wastewater Treatment Plant: Ideally, the SAT site should be within 5 km of the wastewater treatment plant to minimize costs associated with the transport of effluent. Distances of 5–20 km are moderately suitable but will incur higher conveyance costs. Sites located more than 20 km away are considered unsuitable due to the significant increase in pumping costs and potential logistical complexities [17].

Elevation Difference Relative to Wastewater Treatment Plant: An elevation difference of less than 50 m is preferable to facilitate the gravity flow of wastewater and reduce pumping energy requirements. Elevation differences greater than 50 m are unsuitable as they may necessitate additional energy expenditure for pumping and could complicate the overall design and operation of the SAT system [17].

These site selection criteria are essential to the preliminary assessment phase of SAT system planning, as they directly impact the feasibility, design, cost, and operational efficiency of the treatment system. The proper application of these criteria ensures the selection of a site that is not only technically appropriate but also economically viable for sustainable wastewater management.

5.2 Required Site Area

The total land required for a land treatment system extends beyond just the area needed for actual wastewater treatment. It includes space for buffer zones, facilities for preapplication treatment, storage areas, access roads, pumping stations, and structures for maintenance and administration. Additionally, considerations for environmental sensitivity must be factored into the land allocation. The future growth patterns in the study area and the accessibility of the land treatment site may necessitate reserving additional land for potential expansion or to handle emergencies [30].

To estimate preliminary site area requirements, factors such as wastewater flows, storage needs, and preliminary loading rates are taken into account. The relationship between the treatment field area, the loading rates, and the operating period is critical in this estimation. This relationship is quantitatively expressed in Eq. 2 [33], which is provided in both metric and English standard units, allowing for a comprehensive and adaptable approach to determining the spatial needs of a land treatment system.

$$F = 3.65\frac{Q}{LP}(metric)$$

$$F = 13.443\frac{Q}{LP}(U.S.customary) \quad (2)$$

where, F: field area, ha (acres); Q: Average flow, m^3/d (mgd); L: loading rate, cm/wk (in/wk); P: period of application, wk/yr.

6 Examples of Soil-Aquifer Treatment (SAT) Applications

Some case studies on SAT application highlight its effectiveness in various settings.

In Egypt, Amin et al. [21] conducted a study on an SAT system located west of Sohag. Their research focused on assessing the system's capability in treating wastewater for reuse. The findings revealed that the Sohag SAT system markedly enhanced water quality, significantly improving various parameters like pH, dissolved oxygen (DO), BOD, chemical oxygen demand (COD), total suspended solids (TSS), and heavy metals. Notably, the system achieved high removal efficiencies: 98.3% for BOD, 91% for COD, and over 99.99% for total and fecal coliforms. Although there was a slight increase in total dissolved solids (TDS) and boron (B) levels, heavy metals were almost entirely eliminated. The study also underscored the system's economic viability, noting that its costs were less than 34% of those associated with tertiary treatment. Given its affordability and significant environmental and public health benefits, the SAT system is presented as a practical solution for unrestricted irrigation and safe wastewater reuse in Egypt.

In Egypt also, Hani et al. [42] created a map identifying potential sites for MAR, including SAT using reclaimed water, in the West Delta. This was achieved by integrating multi-criteria decision analysis with remote sensing and GIS techniques. The analysis identified optimal areas for reclaimed water infiltration in the northeast part of the region. The results of this analysis aligned well with field data gathered from existing MAR sites, confirming the validity and effectiveness of the approach.

In Kuwait, Akber et al. [43] conducted a study to evaluate the long-term effectiveness of the SAT technique in enhancing the quality of treated wastewater in Kuwait's prevailing conditions. The findings revealed that despite the soil's low clay content, significant improvements in the quality of the tertiary treated wastewater were observed. Notable reductions were achieved in the levels of ammonia (>90%), iron (>80%), organic carbon (>90%), biological oxygen demand (BOD) (100%), and bacteria (ranging from 50 to 100%, depending on the type). The study also brought to light that soil leaching in the SAT process tends to elevate the total dissolved solids in the water that has percolated through. Furthermore, it was observed that the nitrification process inherent in SAT leads to an increase in the nitrate levels in the water over an extended period.

In Iran, Mahmoudi et al. [44] evaluated the ability of soil in the arid Varamin region to naturally attenuate inorganic components found in Tehran City's treated municipal wastewater. They created a simulated SAT pond using four columns, each 30 cm tall and 4 cm in diameter, filled with sandy loam soil sourced from an artificial recharge pond in the Varamin plain. These columns were irrigated with secondary treated wastewater from the Shahre-Rey treatment plant, following a cycle of 12 h of wetting and drying. The study found that nitrate removal efficiency during the experiment was approximately 18.4%. Phosphate removal was more effective, with an average removal rate of 83%. Post-experiment soil column analysis revealed that phosphorus primarily accumulated in the top 5 cm of the soil column.

In Jerba Island, South Tunisia, Anane et al. [45] integrated a single-objective Analytic Hierarchy Process (AHP) method with a GIS and field studies to assess the suitability of SAT. The evaluation included various criteria encompassing technical, environmental, and cost factors. The findings revealed that a substantial area, about 1489 hectares, is appropriate for SAT implementation. This area is larger than what is needed to infiltrate the total volume of treated wastewater from all wastewater plants on the island. Located primarily within agricultural regions and close to coastlines, these potential SAT sites could enhance groundwater availability for irrigation purposes and help alleviate the marine intrusion issues plaguing the Jerba aquifer. The most favorable sites for SAT were identified nearer to the Aghir treatment plant compared to other wastewater treatment plants.

In India, Nema et al. [46] conducted an evaluation of the SAT method using primary treated wastewater in the Ahmedabad region. The SAT system established there consisted of two primary settling basins, two infiltration basins, and two production wells situated at the center of the infiltration basins for drawing out reclaimed wastewater. The findings indicated that the SAT system was more efficient and economical compared to traditional wastewater treatment methods, advocating its broader implementation. Additionally, the study projected that the adoption of SAT technology could evolve into a straightforward and financially viable wastewater treatment option for developing countries.

In Greece, Pliakas et al. [47] focused on designing an effective SAT system for groundwater recharge. This design was based on an analysis of the geological and hydrogeological conditions of the coastal aquifer at northeastern Greece. Their findings indicated that groundwater levels, which were higher than the mean sea level due to the recharge dynamics of the aquifer, effectively reduced the risk of seawater intrusion.

in Arizona, USA, Fox et al. [48] explored the viability of SAT systems across five different soils, using various sources of reclaimed water in soil column experiments. Four of these soils were sourced from a potential SAT site near Phoenix, Arizona, USA, and were identified as Agua Fria sand, North Pond silt, South Pond silt, and agricultural field clay. The fifth soil came from the Sweetwater SAT site in Tucson, Arizona. These soils were chosen for their diverse range of properties. The study demonstrated that soil aquifer treatment is effective in removing organic carbon, without leading to significant accumulation or breakthrough due to adsorption.

7 Limitations and Challenges

The application of SAT systems in the MENA region presents a variety of challenges, which stem largely from the unique climatic, geological, and wastewater quality characteristics of the area. Grinshpan et al. [49] outline the following key limitations and risks associated with implementing SAT:

1. High soil permeability, ideal in fine sand, loamy sand, and sandy loam soils [50].
2. A vadose zone depth of not less than 2 m [30], with a preference for >5 m [17], and ideally ranging from 10 to 30 m [51]
3. The requirement for large areas of land to establish infiltration basins, a condition that may be challenging to meet, especially in urban settings [17].

1. Reduced Winter Infiltration Rates: Factors affecting SAT basin infiltration rates in winter include:
   - Lower microbial activity due to cooler temperatures, leading to diminished organic matter decomposition and accumulation in top soil layers, impeding infiltration [52].
   - Increased water viscosity with decreasing temperature, e.g., a rise in water viscosity by over 30% when temperature drops from 25 to 10 °C, affecting soil hydraulic conductivity and reducing infiltration rates [52–54].
2. Impact of Shorter Drying Periods: Consequences of reduced drying periods in SAT basins include.
   - Increased hydraulic load due to higher wastewater inflow and stormwater transmission, leading to shorter drying intervals [55–57].
   - Observations of inadequate air transfer and ensuing anaerobic conditions during short dry periods, affecting basin efficiency [58].

However, for optimal functionality of SAT systems in the MENA region, it's crucial to address these challenges:

- Groundwater Contamination Risk: If wastewater isn't properly treated before percolation, there's a significant risk of contaminating groundwater. This becomes a more pressing issue in areas with lenient or challenging-to-enforce environmental regulations.

- Need for Regulatory Support: The effective use of SAT systems hinges on having robust regulatory frameworks in place. These frameworks should focus on environmental protection, quality control, and the sustainable management of groundwater resources. However, the development and enforcement of such regulations can be a complex task in some countries of the MENA region.
- Technical Expertise and Adaptation to Local Conditions: Implementing and managing SAT systems requires advanced technical skills. Moreover, adapting SAT technology to the unique environmental and climatic conditions of the region calls for specialized research and development efforts.
- Groundwater Monitoring: Continuous and detailed monitoring of groundwater is essential. This vigilance is necessary to prevent any decline in water quality, ensuring the long-term success of SAT systems.

8 Conclusions

Soil aquifer treatment (SAT) is gaining prominence as a viable solution to global wastewater reclamation. The utilization of partially treated wastewater into SAT systems for groundwater recharge and recovery plays a crucial role in the treatment and conservation of wastewater for future purposes. The effectiveness of SAT can be influenced by factors such as site attributes, the initial wastewater quality, climatic conditions, hydraulic loading rates, and system design.

Strategically managing the application of partially treated wastewater to soils can substantially reduce contaminant concentrations. The utilization of reclaimed water also holds promise as a new irrigation source, especially beneficial in arid and semi-arid regions like the MEAN region. To gauge the effectiveness of SAT in removing pollutants from partially treated wastewater, it is essential to examine the water quality at two key stages: before entering the SAT system and after undergoing SAT treatment, as observed in the outflow from wells. Furthermore, SAT systems should include monitoring wells, sensors, and control mechanisms for ongoing assessment of water quality, flow rates, and system performance. This real-time data enables operators to make timely adjustments to enhance treatment processes and ensure adherence to regulatory standards. These assessments indicate a significant potential for improving the quality of partially treated wastewater through the SAT process.

9 Recommendations

Addressing the challenges faced by the MENA region in applying SAT systems is crucial for sustainable water management. The following recommendations aim to mitigate the difficulties arising from climatic, environmental, socioeconomic, and geopolitical factors.

Development of SAT systems tailored to local environmental conditions. This includes designing systems that account for soil characteristics, climate, and available water resources.

Implementation of water conservation strategies to maximize available water for SAT systems. Establishment of strict guidelines and monitoring systems to prevent groundwater contamination. Regular testing and quality control measures should be integral to the SAT process.

Additionally, establishing a dedicated program for specialized research and development in SAT technology is recommended. This program should specifically consider and incorporate the unique environmental and climatic conditions of the region.

References

1. Mekonnen MM, Hoekstra AY (2016) Four billion people facing severe water scarcity. Sci Adv 2(2):e1500323. https://doi.org/10.1126/sciadv.1500323
2. Kalogirou SA (2005) Seawater desalination using renewable energy sources. Prog Energy Combust Sci 31(3):242–281
3. Taft HL (2015) Water scarcity: global challenges for agriculture. In *Food, energy, and water*, pp 395–429. https://doi.org/10.1016/B978-0-12-800211-7.00016-8
4. Khan MA, Rehman S, Al-Sulaiman FA (2018) A hybrid renewable energy system as a potential energy source for water desalination using reverse osmosis: a review. Renew Sustain Energy Rev 97:456–477
5. Gleick PH, Cain NL, Haasz D, Henges-Jeck C, Hunt C, Kiparsky M, Moench M, Palaniappan M, Srinivasan V, Wolff GH (2004) The world's water 2004–2005: the biennial report on freshwater resources. Island Press
6. Gleick PH, Wolff GH, Cooley H, Palaniappan M, Samulon A, Lee E, Morrison J, Katz D (2013) The World's Water 2006–2007: the biennial report on freshwater resources. Island Press
7. de Waal D, Khemani S, Barone A, Borgomeo E (2013) The economics of water scarcity in the Middle East and North Africa: institutional solutions. World Bank Publications
8. El-Rawy M, Batelaan O, Al-Arifi N, Alotaibi A, Abdalla F, Gabr ME (2023) Climate change impacts on water resources in arid and semi-arid regions: a case study in Saudi Arabia. Water 15(3):606. https://doi.org/10.3390/w15030606
9. El-Rawy M, Fathi H (2023) Groundwater pollution sources and its quality in the Kingdom of Saudi Arabia: state of the art. In: The handbook of environmental chemistry. Springer, Berlin, Heidelberg. https://doi.org/10.1007/698_2023_1050
10. El-Rawy M, Batelaan O, Alshehri F, Almadani S, Ahmed MS, Elbeltagi A (2023) An integrated GIS and machine-learning technique for groundwater quality assessment and prediction in Southern Saudi Arabia. Water 15(13):2448. https://doi.org/10.3390/w15132448
11. El-Rawy M, Fathi H, Abdalla F, Alshehri F, Eldeeb H (2023) An Integrated principal component and hierarchical cluster analysis approach for groundwater quality assessment in Jazan, Saudi Arabia. Water 15(8):1466. https://doi.org/10.3390/w15081466

12. Westcot DW (1997) Quality control of wastewater for irrigated crop production (Water reports-10). Food and Agriculture Organization of the United Nations, Rome, Italy
13. Batarseh MI, Rawajfeh A, Ioannis KK, Prodromos KH (2011) Treated municipal wastewater irrigation impact on olive trees (Olea Europaea L.) at Al-Tafilah, Jordan. Water Air Soil Pollut 217:185–196
14. El-Rawy M, Abdalla F, El Alfy M (2020) Water Resources in Egypt. In: Hamimi Z, El-Barkooky A, Martínez Frías J, Fritz H, Abd El-Rahman Y (eds) The geology of Egypt. Regional geology reviews. Springer, Cham. https://doi.org/10.1007/978-3-030-15265-9_18
15. El-Rawy M, Zlotnik VA, Al-Raggad M, Al-Maktoumi A, Kacimov A, Abdalla O (2016) Conjunctive use of groundwater and surface water resources with aquifer recharge by treated wastewater: evaluation of management scenarios in the Zarqa River Basin, Jordan. Environ Earth Sci 75:1146. https://doi.org/10.1007/s12665-016-5946-1
16. El-Rawy M, Al-Maktoumi A, Zekri S, Abdalla O, Al-Abri R (2019) Hydrological and economic feasibility of mitigating a stressed coastal aquifer using managed aquifer recharge: a case study of Jamma aquifer, Oman. J Arid Land 11:148–159. https://doi.org/10.1007/s40333-019-0093-7
17. Sharma SK, Kennedy MD (2017) Soil aquifer treatment for wastewater treatment and reuse. Int Biodeterior Biodegrad 119:671–677. https://doi.org/10.1016/j.ibiod.2016.09.013
18. Rüetschi D (2020) DOC-removal in recharge water by soil-aquifer-treatment. In: Groundwater. CRC Press, pp 279–280
19. Dillon P (2005) Future management of aquifer recharge. Hydrogeol J 13:313–316. https://doi.org/10.1007/s10040-004-0413-6
20. Donn M, Reed D, Vanderzalm J, Page D (2020) Assessment of *E. coli* attenuation during infiltration of treated wastewater: a pathway to future managed aquifer recharge. Water 12(1):173. https://doi.org/10.3390/w12010173
21. Amin HM, Gad AA, El-Rawy M, Abdelghany UA, Sadeek RA (2021) Assessment of wastewater contaminant concentration through the vadose zone in a soil aquifer treatment system. Appl Ecol Environ Res 19(3):2385–2403
22. Asano T (ed) (1998) Wastewater reclamation and reuse: water quality management library, vol 10. CRC Press
23. Bixio D, Thoeye C, De Koning J, Joksimovic D, Savic D, Wintgens T, Melin T (2006) Wastewater reuse in Europe. Desalination 187(1–3):89–101. https://doi.org/10.1016/j.desal.2005.04.070
24. Morrison CM, Betancourt WQ, Quintanar DR, Lopez GU, Pepper IL, Gerba CP (2020) Potential indicators of virus transport and removal during soil aquifer treatment of treated wastewater effluent. Water Res 177:115812. https://doi.org/10.1016/j.watres.2020.115812
25. Shabani F, Aflaki R, Minamide T, Venezia T, Stenstrom MK (2020) Soil aquifer treatment to meet reclaimed water requirements. Water Environ Res: Res Publ Water Environ Federation 92(2):266–277. https://doi.org/10.1002/wer.1245
26. Idelovitch E (1978) Wastewater reuse by biological-chemical treatment and groundwater recharge. J (Water Pollution Control Federation) 2723–2740. https://www.jstor.org/stable/25040213
27. Shoushtarian F, Negahban-Azar M (2020) Worldwide regulations and guidelines for agricultural water reuse: a critical review. Water 12(4):971. https://doi.org/10.3390/w12040971
28. Liang X, Zhan H, Zhang YK (2018) Aquifer recharge using a vadose zone infiltration well. Water Resour Res 54(11):8847–8863. https://doi.org/10.1029/2018WR023409
29. Idelovitch E (1981) Unrestricted irrigation with municipal wastewater. In: Environmental engineering. ASCE, pp 243–256
30. Smith JE (2006) Process design manual: land treatment of municipal wastewater effluents. EPA/625/R-06/016. US Environmental Protection Agency 193
31. Tchobanoglus G, Burton F, Stensel HD (2003) Wastewater engineering: treatment and reuse. Am Water Works Assoc J 95(5):201
32. Smith JW, Crites RW (2001) Rational method for the design of organic loading rates in a land application system. In: WEFTEC 2001. Water Environment Federation, pp 1–9

33. Hossner LR, Weaver RW, Dronen NO, Fehrmann RC, Foster BG, Heck FC, Kao CW, Wagggoner JA (1978) Sewage disposal on agricultural soils: chemical and microbiological implications. Environmental Protection Agency, Office of Research and Development, Robert S. Kerr Environmental Research Laboratory
34. Pound CE, Crites RW, Olson JV (1978) Long-term effects of land application of domestic wastewater: Hollister, California, Rapid Infiltration Site. Environmental Protection Agency, Office of Research and Development, Robert S. Kerr Environmental Research Laboratory
35. Crites R, Tchobanoglous G (1998) Small and decentralized wastewater management systems. McGraw Hill Co., New York, NY
36. Page AL (2018) Land application of sludge. CRC Press
37. Reed SC, Crites RW, Middlebrooks EJ (1995) Natural systems for waste management and treatment, 2 edn. McGraw-Hill, Inc.
38. Sposito G (2008) The chemistry of soils. Oxford University Press
39. Brady NC, Weil RR, Weil RR (2008) The nature and properties of soils, vol 13. Prentice Hall, Upper Saddle River, NJ, pp 662–710
40. Sparks DL, Singh B, Siebecker MG (2022) Environmental soil chemistry. Elsevier
41. Joshaghani M, Ghasemi-Fare O (2021) Exploring the effects of temperature on intrinsic permeability and void ratio alteration through temperature-controlled experiments. Eng Geol 293:106299. https://doi.org/10.1016/j.enggeo.2021.106299
42. Hani HM, El Din MM, Khalifa A, Elalfy E (2023) Development of suitability map for managed aquifer recharge: case study, West Delta, Egypt. AQUA-Water Infrastruct Ecosyst Soc 72(6):868–884. https://doi.org/10.2166/aqua.2023.177
43. Akber A, Mukhopadhyay A, Al-Senafy M, Al-Haddad A, Al-Awadi E, Al-Qallaf H (2008) Feasibility of long-term irrigation as a treatment method for municipal wastewater using natural soil in Kuwait. Agricult Water Manag 95(3):233–242. https://doi.org/10.1016/j.agwat.2007.09.015
44. Mahmoudi N, Nakhaei M, Porhemmat J (2018) Assessment of an arid region soil capacity on natural attenuation of municipal treated wastewater: a column experiment using soil of Varamin area, Iran. Environ Earth Sci 77:1–4. https://doi.org/10.1007/s12665-018-7292-y
45. Anane M, Kallali H, Jellali S, Ouessar M (2008) Ranking suitable sites for Soil Aquifer Treatment in Jerba Island (Tunisia) using remote sensing, GIS and AHP-multicriteria decision analysis. Int J Water 4(1–2):121–135. https://doi.org/10.1504/IJW.2008.018151
46. Nema P, Ojha CS, Kumar A, Khanna P (2001) Techno-economic evaluation of soil-aquifer treatment using primary effluent at Ahmedabad, India. Water Res 35(9):2179–2190. https://doi.org/10.1016/S0043-1354(00)00493-0
47. Pliakas F, Kallioras A, Diamantis I, Stergiou M (2011) Groundwater recharge using a Soil Aquifer Treatment (SAT) system in NE Greece. In: Lambrakis N, Stournaras G, Katsanou K (eds) Advances in the research of aquatic environment. Environmental earth sciences. Springer, Berlin, Heidelberg. https://doi.org/10.1007/978-3-642-19902-8_34
48. Fox P, Aboshanp W, Alsamadi B (2005) Analysis of soils to demonstrate sustained organic carbon removal during soil aquifer treatment. J Environ Qual 34(1):156–163. https://doi.org/10.2134/jeq2005.0156a
49. Grinshpan M, Furman A, Dahlke HE, Raveh E, Weisbrod N (2021) From managed aquifer recharge to soil aquifer treatment on agricultural soils: concepts and challenges. Agric Water Manag 255:106991. https://doi.org/10.1016/j.agwat.2021.106991
50. Pescod MB (1992) Wastewater treatment and use in agriculture—FAO irrigation and drainage. Food and Agriculture Organization of the United Nations, Rome. https://doi.org/10.M-56
51. Aharoni A, Guttman J, Cikurel H, Sharma S (2011) Guidelines for design, operation and maintenance of SAT (And Hybrid SAT) systems. MEKOROT and UNESCO-IHE. Deliverable number 3.2.1.f-i, EU SWITCH Project. http://www.switchurbanwater.eu/outputs/pdfs/W3-2_GEN_RPT_D3.2.1f-i_Guidelines_for_ design_of_SAT_systems.pdf
52. Nadav I, Arye G, Tarchitzky J, Chen Y (2012) Enhanced infiltration regime for treated-wastewater purification in soil aquifer treatment (SAT). J Hydrol 420–421:275–283. https://doi.org/10.1016/j.jhydrol.2011.12.013

53. Hillel D (1980) Fundamentals of soil physics, 1st edn. Academic Press, New York
54. Casanova J, Devau N, Pettenati M (2016) Managed aquifer recharge: an overview of issues and options. In: Jakeman AJ, Barreteau O, Hunt RJ, Rinaudo JD, Ross A (eds) Integrated groundwater management: concepts, approaches and challenges. Springer International Publishing, pp 413–434. https://doi.org/10.1007/978-3-319-23576-9
55. Joannis C, Commaille J, Dupasquier B (2002) Assessing infiltration flow-rates into sewers. In: Global Solutions for Urban Drainage 2002, pp 1–15. https://doi.org/10.1061/40644
56. De Bénédittis J, Bertrand-Krajewski JL (2005) Infiltration in sewer systems: comparison of measurement methods. Water Sci Technol 52:219–227. https://doi.org/10.2166/wst.2005.0079
57. Negev I, Shechter T, Shtrasler L, Rozenbach H, Livne A (2020) The effect of soil tillage equipment on the recharge capacity of infiltration ponds. Water 12:1–11. https://doi.org/10.3390/w12020541
58. Ben Moshe S, Weisbrod N, Barquero F, Sallwey J, Orgad O, Furman A (2020) On the role of operational dynamics in biogeochemical efficiency of a soil aquifer treatment system. Hydrol Earth Syst Sci 24:417–426. https://doi.org/10.5194/hess-24-417-2020

Applicability of River Bank Filtration for Affordable Drinking Water in the MENA Region: Hydrogeological Insights and Water Quality Implications

Ahmed Abdelrady, Ali A. A. Obeid, and C. Belloni

Abstract This chapter delves into the feasibility of implementing River Bank Filtration (RBF) as a sustainable and cost-effective solution for drinking water supply in Middle Eastern countries. Comprehensive assessments across diverse hydrogeological settings highlight both positive and negative influences on water quality. Notably, the study shows the successful application of RBF in alluvial aquifers along the Nile River (Egypt), Zarqa River (Jordan), and Sirwan Al-Kufa Rivers (Iraq), affirming the technique's viability. The interaction between surface water systems and adjacent coarse aquifers underscores the long-term effectiveness of RBF. The technology remarkably efficiently removes various pollutants, including organic and microorganic contaminants. However, careful design considerations are essential to minimize the contribution of ambient groundwater, known for elevated concentrations of organic compounds and metals. The ongoing operation of RBF wells is critical to sustained success, which helps increase the share of bank filtrate and counteracts declining surface water levels. Strategic well placement near the surface water system (aerobic zone) is recommended to prevent mobilization of iron (Fe) and manganese (Mn) into the bank filtrate. Economic analyses underscore RBF as a cost-effective alternative, delivering high water quality while surpassing conventional treatment techniques.

Keywords Bank filtration (PBF) · Managed aquifer · Groundwater management · Hydrogeological · Water quality · MENA

A. Abdelrady (✉) · C. Belloni
Wetsus, European Centre of Excellence for Sustainable Water Technology, Oostergoweg 9, 8911, MA Leeuwarden, The Netherlands
e-mail: Ahmed.Mahmoud@wetsus.nl

C. Belloni
e-mail: carlo.belloni@wetsus.nl

A. A. A. Obeid
Institute of Hydraulic Engineering and Water Resources Management, Vienna University of Technology, Vienna, Austria
e-mail: ali.obeid@tuwien.ac.at

M. El-Rawy and A. Negm (eds.), *Managed Aquifer Recharge in MENA Countries*, Earth and Environmental Sciences Library,
https://doi.org/10.1007/978-3-031-58764-1_5

1 Introduction

River bank filtration (RBF) is a natural treatment technique that produces high water quality at a relatively low cost. RBF draws water from hydraulically connected wells to a surface water resource. RBF is a process that exposes surface water to subsurface flow caused by lowering the hydraulic head prior to extraction from vertical or horizontal wells. As a result of pumping, a pressure head gradient is established between the river and the wells that induce the water to flow from the surface water system towards the RBF wells provided that there are no low hydraulic conductivity layers such as clay, artificial or natural barriers to this subterranean flow exist. The pumped water is a mixture of the infiltrated water and ambient groundwater. Following World War II (WWII), Europe's surface water systems faced significant pollution from industrial and municipal waste. Only Riverbank Filtration (RBF) emerged as a solution capable of delivering consistently high water quality [1]. In recent times, there has been a notable extension in the utilization of this methodology to regions characterized by disparate climatic and environmental profiles [2]. This chapter undertakes an examination of the viability of deploying this sustainable technique within the Middle Eastern domain.

RBF can improve water quality by lowering turbidity, microbial contaminants, microcystins, pathogens, heavy metals, organic matter, and inorganic contaminants like phosphorus and nitrate [3]. The RBF technique has a high potential for removing these dissolved pollutants that are difficult to remove in surface water treatment plants, as it reduces dissolved organic matter (DOM) and disinfection byproduct precursors by 50%. Pollutant attenuation is based on processes such as biodegradation, adsorption, precipitation, filtration, and mixing/dispersion that occur during the flow of water from the surface water system to the pumped wells [4].

The effectiveness of this technique is determined by a variety of factors, including aquifer geology, aquifer shape, surface water quality, organic matter types in surface and groundwater environments, river bed composition and clogging, and land use in the local catchment area [5]. Alluvial aquifers stand out as prime locations due to their abundant production potential, strong connection to surface water networks, and close proximity to areas of demand. RBF can provide high water quality under extreme environmental conditions such as surface water quality variation, chemical load shock, and large temperature variations. Furthermore, the flow of surface water through the (river or lake) banks reduces contaminant concentrations in the surface water and acts as a barrier against concentration peaks caused by accidental pollutant spills. Another advantage of this technique for the environment is that no waste is produced during its operation [6].

Natural treatment systems, such as RBF, are critical for developing and developed ones. For more than 140 years, it has been widely used as a water treatment technique in developed countries. There are over 100 RBF sites in Europe that produce high-quality water. Many of them have been in operation for over than a century [1]. RBF contributes approximately 7% of the national drinking water supply in the Netherlands [7]. It supplies approximately 8% of Germany's drinking water. Duesseldorf, located on the Rhine, is completely supplied with RBF-treated drinking water. More than 20 million people in the Rhine basin receive drinking water that is directly or indirectly derived from river water, mostly via RBF. This technique is primarily used as a pretreatment step for microorganisms in North America [8]. RBF wells are operated in European and North American countries in favourable environmental conditions, such as high-quality source water, by people with extensive experience designing RBF systems.

This technique has not been fully utilized in developing countries. Many countries have recently studied the feasibility of implementing this technology under their own local conditions [9]. There are now over 80 RBF sites along the Nile River in Egypt. In this chapter, we discuss the feasibility of implementing this sustainable technique in MENA countries, the technique's ability to remove chemical and biological contaminants under arid climate conditions, and we finally propose some guidelines for the successful application of the technology in the MENA region.

2 Experience of River Bank Filtration

RBF has been used for many years as a multi-objective natural treatment technique that removes most surface water pollutants (even those that conventional water treatment does not remove) and produces high water quality at a low cost. It can be used to replace or supplement current water treatment techniques by providing a strong barrier and lowering treatment costs. RBF is distinguished by the avoidance or reduction of the use of chemical disinfectants in order to produce biologically stable water. Another advantage of RBF is that it may be used as a water storage capacity in regions with seasonal variations in precipitation and run-off regimes (e.g. monsoon, floods, and droughts). Moreover, mixing bank filtrate with ambient groundwater increases the groundwater supply and dilutes contaminants.

Many water utilities in developed countries have used the RBF technique to produce drinking water. RBF is used primarily along the Rhine in the region between the tributaries Sieg and Ruhr, on the Elbe between Dresden and Torgau, and in the Berlin area. It is used along the Rhine and Meuse rivers in the Netherlands. It is also applied in many other European countries e.g. Austria, Switzerland and France. The rivers quality began to deteriorate in recent years, and an anaerobic environment developed in the infiltration area, as well as a high reduction of iron and manganese. Consequently, the implementation of subsequent remedial interventions has become imperative to elevate the water quality of pumped water within the contemporary era. Consequently, post-treatment measures have become requisite. While raw water quality has exhibited improvement in recent years, it is imperative to consider the periodic fluctuations in river water quality and hydraulic dynamics influenced by climatic conditions. In concert with ozonation and filtration techniques, the utilization of granular activated carbon has been adopted to augment the treatment of bank filtrate, facilitating the elimination of chemical contaminants [10].

Due to its high production capacity, horizontal wells were first used to produce water for industrial use and only one-third of the first developed horizontal wells were used for drinking purposes. In the last years, this design has spread to produce drinking water in large quantities and of the right quality [11].

3 Bank Filtration Process

RBF has the potential to significantly reduce the concentrations of a wide range of chemical and biological contaminants, including turbidity, dissolved chemicals, natural organic matter, and microbes. The surface water system's bed and bank act as a barrier to attenuate surface water pollutants. Filtration, biodegradation, reduction and oxidation, adsorption–desorption, chemical precipitation, and dilution are some of the mechanisms used to remove pollutants during RBF (Fig. 1). Temperature, redox environment, distance to the river, and raw water quality are all factors that influence the infiltration process and thus RBF performance. The majority of the contaminants are removed within the first few centimetres of the river bed, while others are removed further down the infiltration path. However, persistent contaminants may migrate closer to the RBF wells. As a result, the infiltration zone can be divided into two major zones [5].

1. Colmation layer: it is a highly active biological zone where intensive biodegradation and adsorption processes take place within a short travel time. Experiments have proven that most treatment processes take place in this layer.

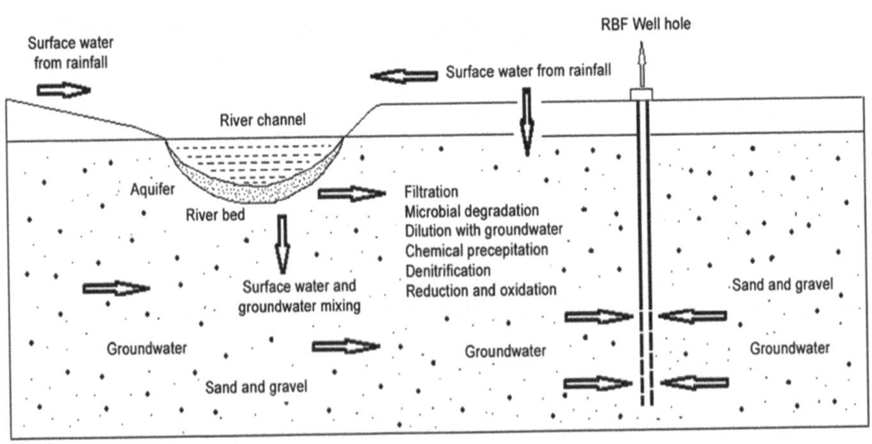

Fig. 1 Bank filtration processes [12]

2. The successive subsurface passage: The subsequent subsurface transition zone is characterized by diminished biological and sorption activities, wherein the predominant processes involve mixing with ambient groundwater and the prevailing dilution effect.

The oxygen in the surface water is almost completely consumed by aerobic microorganisms degrading organic matter, resulting in an anaerobic infiltration environment. The undesirable effects of switching from an oxidizing to a reducing environment include heavy metal dissolution and increased ammonia, calcium, and magnesium concentrations, as well as the formation of hydrogen sulphide and other odorous sulphur compounds [13].

4 Siting and Design of RBF Wells

RBF discharges water from wells close to the surface water system. As a result, the characteristics of the surface water and the adjacent aquifer have a large impact on the processes occurring at the infiltration area and thus on the quantity and quality of bank filtrate. A good site selection based on surface water and groundwater characteristics and their interaction is a critical step in designing RBF wells and transferring this technique to developing countries [14]. Srisuk et al. [15] developed a method to select appropriate RBF wells sites in Thailand. The selection procedure was divided into 3 Sequential steps: (1) Identify the RBF potential areas in the river basin; (2) Identify potential local areas within the potential river basins; (3) Identify the RBF potential sites and locations within the local areas. For each step, different criteria had been identified by the author to be achieved in order to decide the feasibility of the site for RBF application. Alternatively, Mohd et al. [16] used a combination of GeoEye-1 satellite image and a GIS tool to develop a site suitability map for RBF locations in Dengkil (Malaysia). The satellite image classification was performed to differentiate 6 major land use classes in the study area. Range-land and barren land were selected by the author as the most suitable sites for RBF well-sitting. The GIS tool was used to compare the potential sites based on two criteria: (1) Distance from the river should be within the range of 10–20% of river width; (2) The RBF well location should be at least 15 m from the built-up area. The results showed that the selected sites are in agreement with the existing borehole location of the study area.

4.1 Site Selection Parameters

RBF has an advantage over slow sand filtration in that it produces higher clean water quantities. The RBF yield is determined by several interconnected factors. The main site-specific parameters that should be considered during RBF well design and installation are aquifer characteristics and river hydrology. Furthermore, complex processes such as clogging that occur during RBF well operation have a significant impact on RBF yield by lowering the hydraulic conductance of the surface water bed and thus reducing the recharge rate to the aquifer. Clogging is reduced by the river's high bed load transport, which removes deposits from the river bed [17].

4.2 *Aquifer Characteristics*

Most RBF wells are installed in alluvial aquifers which are hydraulically connected to the surface water system. Other geological settings, such as karst or bedrock formations, are less effective than alluvial deposits in terms of providing high-quality and quantity of water. An alluvial aquifer is mainly composed of sand, gravel, clay, silt, large cobbles and boulders. These consolidated materials are able to create a tortuous pathway during infiltration, which promotes microbial growth and enhances biodegradation, chemical transformation, and physical straining of impurities, thus improving bank filtrate quality. However, alluvial aquifers with clay overbank layer larger than 20 mbgs and low hydraulic conductivity should be avoided [18].

Grain size and soil texture are important parameters that affect RBF performance. Grain size determines the transmissivity and hydraulic conductivity of the aquifer and thus affects bank filtrate quantity. Coarse-grained alluvial materials with high hydraulic conductivity are most suitable for RBF systems. In general, poorly sorted granular material is preferred over well-sorted granular materials as it provides a higher degree of natural filtration. Soil texture also plays a vital role in determining the bank filtrate quality. The soil should exhibit filtration capacity to be considered. For example, limestone has a high ability to increase the flow velocity and shorten the travel time and so decreasing the filtration process efficiency [19].

Groundwater transmissivity and the recharge infiltration rate determine the bank filtrate quantity. The volume of the aquifer is a function of the saturated thickness, width, and length of the aquifer. Length can be neglected as the alluvial aquifer extended along the surface water system. On contrary, alluvial aquifer depth should be considered as it affects the bank filtrate availability and the travel time. According to Ray et al. [20], the alluvial aquifer thickness should be at least 5 m. Transmissivity is a function of the conductivity of the aquifer medium and the saturated thickness of the aquifer. According to Ray and Shamrukh [21], the hydraulic conductivity of the aquifer should be higher than 10^{-4} m/s.

4.3 *River Characteristics*

The hydrology of the river affects the interaction between the surface and groundwater. A high river discharge rate will shorten the travel time and affect the bank filtrate quality. In contrast, drought and low discharge rates will make the travel time longer and reduce the quantity and quality of the bank filtrate (lowered dilution of contaminants, salinization). Therefore, daily run-off data should be collected and averaged in order to draw the unit hydrograph of the surface water system and to observe the high and low flow regimes together with the period in which those events take place in a year. Runoff regimes affect the RBF yield by altering the river bed through erosion, transport, and deposition thus determining the self-cleaning potential of the river. A river with a consistent hydrograph is characterized by less variability in riverbed depositional and fluvial processes and, thus, less variability in surface water and groundwater interaction. Run-off hydrograph may also be used to propose successful operation procedures. RBF wells can be used instead of conventional surface water treatment when the river flow and discharge rate are high, and thus, the surface water body has a high concentration of suspended solids. In contrast, conventional treatment techniques can be used when the river discharge is low. During low discharge periods, weir structures/systems and inflatable dams can be used to regulate the river flow and prevent the river stage from dropping [22].

River geomorphology is another important characteristic to describe the clogging and erosion process taking place in the river system. Generally, the river can be divided into 3 regions: (1) Upper part with erosion. (2) Middle part with bed load transport. (3) Lower part with deposition. The upper layer is an infeasible location for RBF well installation as the aquifers at these locations tend to be too small to support the application of this technique. The application of this technique is rendered unfeasible as well in the exceedingly low section of the river due to the presence of near-zero slopes and the influence of estuarine dynamics. Most of the RBF field along the Rhine River is located in the middle and lower part [23]. River reaches can be classified into different types based on the morphological units that are located within them. Morphological units include pools, riffles, rapids, plane beds, and cascades. Deyi [24] developed an RBF selection tool based on the type of reach classification (Fig. 2). The author observed that most existing RBF systems are installed on the Transitional Stream (type B) and upper Foothill stream (type C, E, F) which are characterized with width and depth ratio larger than 12.

Islands and meanders are preferred locations for installing RBF wells to produce more bank filtrate quantities, especially if infiltration occurs naturally in the meander. This is possible if the river has a steeper gradient water level than the groundwater in the connected aquifer and has a high hydraulic conductivity. The share of the bank filtrate would be higher if the RBF wells were installed in the meander's curve and lower if the wells were installed outside the meander and further away from the river. On the other hand, wells outside the meander will have a longer travel distance and a greater filtration effect [26].

Old branches that are downstream and are connected to the river exhibit a natural infiltration of river water towards the branch as a result of water level gradients between the river and the branch. Therefore, it is expected that a high proportion of riverbank filtrate can be pumped under these conditions. However, these regions also exhibit high raw water-suspended solids and are often filled with mud. Therefore, bed hydraulic conductivity is very low and the quantity of the bank filtrate is not sustainable [23].

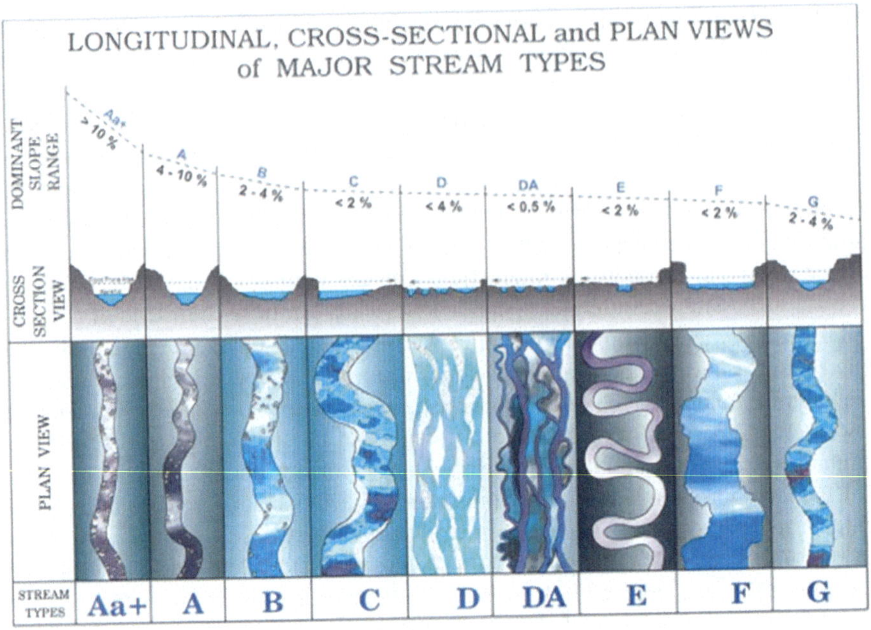

Fig. 2 Longitudinal, cross-sectional, and plan views of major stream types [25]

The river bed has a vital role in determining the quality and quantity of bank filtrate as it is the interface between the river and the aquifer where most of the biodegradation processes take place. The permeability of the river bed is the dominating factor affecting the hydraulic connection between the river and adjacent aquifer and thus affecting the RBF sustainability [27]. An ideal river bed is able to retard microbial pathogen transport and provide optimal water recharge to the infiltration area [20]. River beds consisting of fine particles will have lower conductivities than those composed of coarser particles. Therefore, grain size distribution can be used as a tool to predict the clogging process. The d_{10} size indicates the presence of fine sediments that might increase the clogging development rate. On the other hand, d_{90} size might refer to the presence of an amorphous layer if its portion is large compared to d_{10}. This layer's presence reduces the RBF systems' capacity by protecting trapped sediment from renewal during flood events [28]. In general, areas characterized by erosion and the deposition of very fine particles such as silt and fine sand—such as those found upstream of dams and near the river mouth—are deemed unsuitable for the installation of Riverbank Filtration (RBF) wells and are advised to be avoided [20].

4.4 Clogging Process

Clogging of the river bed can't be avoided during the RBF process. The clogging layer is enriched with heavy metals and adsorbed organic matter. Grischek et al. [17] illustrated that if significant clogging is expected, then RBF wells should be installed close to the bank to ensure that the desired proportion of riverbank filtrate is extracted.

In RBF systems, three types of clogging can be distinguished. The first and most common type of clogging is mechanical clogging. When suspended solids become trapped in the river bed layer during the infiltration process, this type of clogging occurs. As a result, raw water quality may affect the yield of bank filtrate wells by promoting the formation of a clogging (riverbed) layer and decreasing its hydraulic conductivity [29]. Chemical clogging occurs when a large amount of organic matter enters the infiltration area and changes the pH and redox environment, causing precipitation of substances such as $FeCO_3$ in the aquifer pores [30]. Ammonia, nitrate, iron, calcium, and magnesium are the most influential parameters [28]. Third, biological clogging occurs when a biological layer forms on the river's bottom as a result of microbial body accumulation or metabolic by-products. This type of clogging occurs when the raw water contains a sufficient concentration of nutrients. The main factors that influence the formation of this biological layer are dissolved oxygen and total organic carbon (TOC) concentrations in raw water, and thus they can be used as indicators for the biological clogging process [28].

Changes in the physical and chemical properties of the bank filtrate can be used to predict the clogging process. In general, clogging increases total inorganic concentrations (CO_2, HCO_3^+, and CO_3^{-2}), NH_4^+, Fe^{+2} and Mn^{+2}, total dissolved solids, electric conductivity (EC), and sometimes also SiO_2 while decreasing oxidant concentrations (O_2 and NO^{-3} or even SO_4^{+2}). The methodology relies on monitoring fluctuations in electrical conductivity (EC) measurements, juxtaposing them with variations in raw water characteristics, and scrutinizing alterations in the lag time between corresponding peaks. These observations unveil shifts in travel time, potentially indicative of clogging phenomena. Notably, adjustments for temperature variations and their influence on electrical conductivity values must be duly considered in the analysis [7].

Flooding can be beneficial in destroying the clogging layer. However, if the horizontal wells are located near the bank, it may have a negative impact on the quality of the bank filtrate. Destruction of the clogging layer may result in changes in pore pressure and higher flow velocities in the aquifer, increasing the transport rate of dissolved compounds, previously adsorbed particles, bacteria, and viruses, potentially leading to breakthroughs. As a result, moderate floods may be more effective in removing the clogging layer in these situations [17].

4.5 Bank Filtration Wells Design

RBF wells are classified into three types [5], as follows.

A horizontal well (also known as a "collector well") is made up of a large-diameter vertical caisson sunk into the ground and several lateral wells extending out in all directions from the bottom of the vertical well into the alluvial aquifer beneath the river bed. This system has long been used in the United States to produce groundwater and bank filtrate water. This well captures vertical groundwater flow with minimal horizontal flow. Horizontal wells are distinguished by (1) the ability to pump large amounts of water, (2) the ability to expose infiltrated water to more (well) screens, lowering the entrance velocity of the infiltrated water through the well screen and thus minimizing clogging and head loss between the aquifer and the well.

A vertical well constitutes a tubular structure drilled into the saturated layer. The determination of well depth and screen lengths hinges upon factors such as the depth of the saturation layer and the desired pumping rate. Vertical wells predominantly intercept horizontal groundwater flow, with a minor component of vertical flow. This design finds widespread application in pumping groundwater within developing nations. Vertical wells are particularly favored in scenarios necessitating modest water volumes, owing to their straightforward installation process and their associated low costs for installation, operation, and maintenance.

The hydrological characteristics of the aquifer and the hydraulic conditions of the river are used to choose between horizontal and vertical designs. Vertical and horizontal wells can be easily installed at greater depths (for higher production capacities) and farther away from the river in deep and wide aquifers (to increase the degree of filtration). A horizontal design is preferred in shallow and narrow aquifers because screens can be placed at the lowest elevation, maximizing available drawdown. Furthermore, if there is no or only minor clogging of the riverbed, screens may be extended beneath the riverbed to increase the quantity of pumped water. However, the benefits of RBF processes such as mixing and dilution by ambient groundwater, as well as temperature equilibration, may be overlooked.

Vertical well designs have a longer travel time. Therefore, a greater degree of filtration than horizontal well designs. As a result, microorganisms are more likely to break through in horizontal wells. The influence of flood events on nitrate and atrazine breakthrough at an RBF site in Illinois (USA) was studied using pumping rate, well design (vertical and horizontal), and well depth. Large-capacity horizontal wells were found to be particularly vulnerable to nitrate and atrazine breakthrough during flooding, whereas low-capacity vertical wells may not be affected by river water nitrate and atrazine [20].

4.6 Wells Positions and Spacing

The position of the RBF wells and the space between the wells determine the percentage share of bank filtrate in the pumped water as well as the quality characteristics of the bank filtrate. Producing a large amount of water close to a body of surface water will be advantageous. However, rapid clogging may occur due to the high entrance velocities into the riverbed. In contrast, a long distance between the wells and the bank reduces the amount of water pumped. It will, however, improve the quality of the bank filtrate by increasing the removal of bacteria, viruses, and biodegradable organic matter and equilibrating temperature changes. RBF wells can be installed along straight river sections as well as on the inside or outside of river bends. The flow and shear stress of a river is greater on the outer bend than on the inner bend, making it more susceptible to clogging. Well fields with multiple wells spaced apart can be recharged by a long stretch of river. This will help to remove bacteria, viruses, and organic matter [31, 32].

5 Bank Filtrate Quality

The ability of RBF to produce high-quality water depends on various factors, including raw water quality, hydrological characteristics, and geological settings. The hydrological characteristics significantly impact the travel time and redox conditions of the infiltration area, both of which directly impact RBF performance and pumped water quality. Long travel times allow for more sorption and biodegradation processes, which are critical for the removal of chemical and biological contaminants. However, it may have a negative impact by increasing the concentration of dissolved metals in the pumped water. Travel time in the RBF technique is frequently determined by the survival of viruses rather than the removal of organic matter. It was proposed that 60 days in Germany and 70 days in the Netherlands would suffice to obtain biologically stable drinking water. If this criterion is met, no additional treatment is required. The proposed travel time to remove 90% of the microorganic pollutants is more than 6 months. Because bank filtrate is frequently used as pretreated water for conventional drinking water treatment plants in North America, the travel time has been limited to a few hours to a few weeks at most [33].

The redox processes have a significant impact on the quality of bank filtrate. The biochemical processes occurring in the infiltration area, such as organic matter degradation, denitrification, and iron and manganese dissolution, are heavily reliant on the redox environment. Microorganisms degrade organic matter to generate energy by using various acceptor electrons (O_2, NO_3^-, Mn^{+4}, Fe^{+3}, SO_4^{-2}). As a result, reduced species (H_2O, N_2, Mn^{+2}, Fe^{+2}, HS^-) that are detrimental to bank filtrate quality are produced (Fig. 3).

As a result of redox reactions in the infiltration area, toxic contaminants such as arsenate ions may be released into the bank filtrate water. Arsenate ion concentration

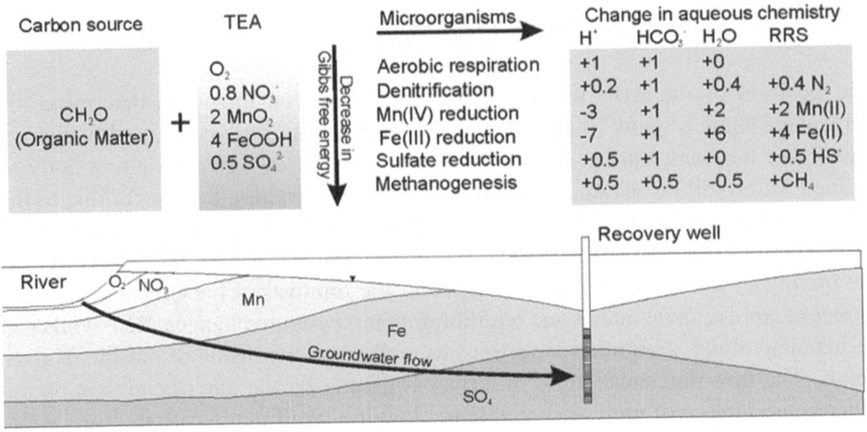

Fig. 3 Redox process in the infiltration area [34]

rises in bank filtrate water due to the oxidation of As-containing sulphide minerals like arsenopyrite. Furthermore, soluble arsenate complexes adsorbed on the iron and manganese surfaces of aquifer materials can be mobilized and dissolved in bank filtrate water under iron and manganese reduction conditions. The As mobilization processes take place primarily in the alluvial aquifer, which is composed primarily of sand and gravel [35, 36].

6 RBF Potential in the MENA Region

Recent studies show that RBF has a high potential for use as a sustainable source of affordable drinking water in many MENA countries, including Egypt, Jordan, and Iraq [37]. In Egypt, the upper layers of quaternary sediments in the Nile Valley and Delta are of particular interest for RBF use. This aquifer is mainly composed of sand and gravel with a high connection to the Nile River (Fig. 4). Shamrukh and Abdel-Wahab [38] illustrated that the Nile aquifer is composed primarily of a thick layer of graded sand and gravel underlain by Pliocene impermeable clays and covered by a silt-clay layer (5–20 m thickness). The thickness of the aquifer ranges from 350 m (at Sohag) to a few meters (at Cairo and Aswan). The aquifer thickness gradually increases north of Cairo until it reaches more than 800 m along the Mediterranean coastline. The aquifer's transmissivity is high, ranging between 5,000 and 20,000 m^2/day. These hydrological features are highly favourable for the use of river bank filtration technology. Therefore, Ghodeif et al. [39] indicated a high potential for RBF along surface waters in the desert fringes of the Nile valley and delta, as well as in the flood plain of Upper Egypt's Nile valley. The author emphasized that RBF wells should be designed to extract water from the aquifer's upper layer in order to increase the amount of bank filtrate. If the clay overbank is 6 m thick and the water

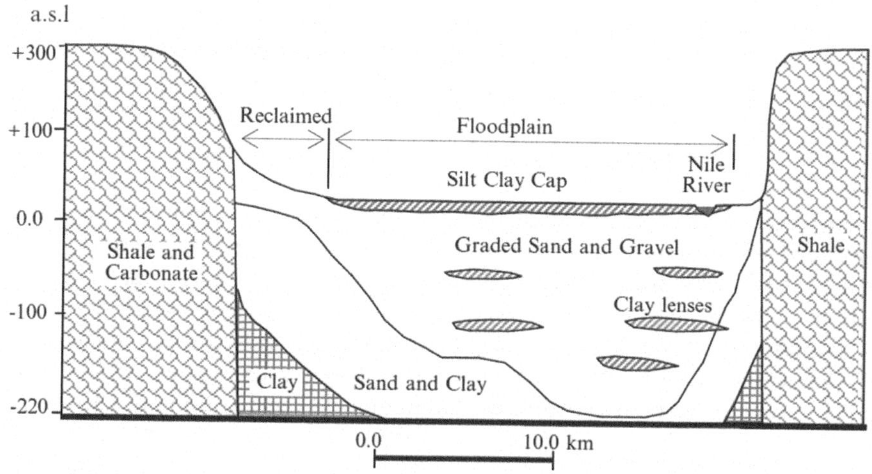

Fig. 4 Nile aquifer general hydrogeological cross section in Nile Valley [37]

table is 4 m below the ground surface (mbgs), the filter screen should be placed between 12 and 24 mbgs.

Bartak [18] conducted a field study to detect the feasibility of applying the RBF technique in Dishna (Qena, Egypt). The authors concluded that the mixing of the contaminated groundwater with the infiltrated water degrades the quality of the pumped water. After 1.5 years of operation, the share of bank filtrate was less than 40%. As a result, the authors emphasized that river leakage is the most critical parameter that should be investigated first in order to determine the feasibility of building an RBF plant in the proposed area.

Abdelrady et al. [34] developed a hydrological model to analyse the performance of RBF systems in Aswan city. The researchers emphasized that design parameters (e.g., number of wells, production capacity, and spacing between wells) should be determined primarily by minimizing the contribution of contaminated groundwater to total pumped water. The Nile River's water level is expected to fall in the near future as a result of either the Renaissance dam in Grand Ethiopia or the effects of climate change. The developed model results revealed that lowering the Nile River level (by 0.5–1.5 m) has a significant effect on the BF parameters (e.g., travel time, the share of bank filtrate) in the wells' initial operation. Nonetheless, operating BF wells for an extended period (approximately 100 days) could mitigate these consequences. This study also revealed that the discharging of untreated wastewater from the septic tanks toward the abstraction well could degrade the quality of the pumped water.

The presence of a high concentration of iron and manganese in the pumped bank filtrate is another drawback that reduces the quality of the bank filtrate and in turn, limits the application of RBF in arid environment countries [11]. Toxic metal(loid)s dissolute in infiltrated water due to the flow of infiltrated water with a high organic matter load and the conversion of the infiltration environment from aerobic to anaerobic. According to Abdelrady et al. [34], infiltrated water with a high concentration of humic compounds can increase the release of Fe, Mn, and As substances from the soil into the infiltrated water. As a result, RBF is more suitable for use in near-surface water systems with lower organic matter concentrations. According to Abdelrady et al. [40], RBF in arid environments has a high capacity to remove biodegradable matter (e.g., biopolymers), but it is less effective against hydrophobic substances like humic acid.

In Jordan, RBF along the Zarqa River is mainly used as a source of irrigation water and has proven its efficacy in producing high-quality drinking water. In terms of annual discharge, the Zarqa River is the second largest river in Jordan, however, it is heavily contaminated with treated domestic and industrial wastewater. According to Blanford et al. [41], the river is highly hydraulically connected to the adjacent alluvial aquifer, and the RBF technique demonstrated a log removal of >4.2 of E. Coli for wells located 5 m from the river (infiltration flow rate = 7 cm/min). However, the pumped water quality still falls short of the standard requirements for drinking water.

RBF demonstrated its efficacy in providing high-quality drinking water along the Sirwan River in Iraq. According to Al-Jaf [42], the pumped bank filtrate is biologically stable and has higher chemical quality than the river and ambient groundwater. Over 90% of the organic matter was removed during infiltration. In contrast, Ghazal et al. [43] evaluated RBF's ability to reduce microbial contamination along the Al-Kufa River. The author noted a reduced microbial content density in bank filtrate wells compared to river water; nevertheless, bank filtrate water remains unsuitable for potable purposes. Concerns arise regarding potential contamination of pumped water due to inadequate design and the proximity of various human and animal activities to the RBF field.

RBF is economically feasible to replace the currently used conventional treatment techniques in the MENA region. RBF units have lower capital and operating costs than conventional water treatment plants [44]. Bartak et al. [45] added that the RBF technique can be used as a pre-treatment step in desalination plants to remove organic matter and reduce biological activity in the water, reducing bio-fouling development in membranes and increasing their sustainability.

7 Conclusions and Recommendations

RBF is a promising and affordable technique to supply water with a high-quality grade for drinking, irrigation, and domestic uses in the MENA region. In this chapter, we highlighted the hydrological condition favourable to applying this sustainable technique, and we define the main prerequisites that could be followed for a successful application, which can be addressed in the following points:

- Riverbank Filtration (RBF) wells are predominantly installed in alluvial aquifers, particularly those intricately linked with the river systems of regions such as Egypt, Iraq, and Jordan. Islands and meanders stand out as prime locales for

RBF well placement, facilitating enhanced production of bank filtrate. However, it should be noted that implementing RBF becomes unfeasible in alluvial aquifers characterized by clay overbank layers exceeding 20 m below ground surface (mbgs) and exhibiting low hydraulic conductivity.

- The design procedures ought to prioritize the diminution of contaminated groundwater within the overall pumped water volume, given its role as the principal source of elevated concentrations of substances such as iron (Fe), manganese (Mn), and organic matter in the pumped water. Nonetheless, the ongoing operation of Riverbank Filtration (RBF) wells holds the potential to elevate the proportion of bank filtrate in the water supply, concurrently ameliorating the impacts stemming from declining surface water levels.
- RBF demonstrated its ability to remove chemical and biological contaminants during infiltration. However, the flow of untreated wastewater from septic tanks into RBF wells degrades the water quality. As a result, governments should take precautions to safeguard the RBF field.
- The main barrier to expanding the use of the RBF technique in the MENA region is the high concentration of metals (e.g., Fe, Mn) in the pumped water. The main mechanism is the anaerobic transformation of the environment and metal mobilization into the infiltrated water. As a result, RBF wells should be installed as close to the river as possible. Further studies are needed to address this issue.

References

1. Eckert P, Irmscher R (2006) Over 130 years of experience with riverbank filtration in Düsseldorf, Germany. J Water Supply: Res Technol—AQUA 55(4):283–291. https://doi.org/10.2166/aqua.2006.0012
2. Abdelrady A, Sharma S, Sefelnasr A, Abogbal A, Kennedy M (2019) Investigating the impact of temperature and organic matter on the removal of selected organic micropollutants during bank filtration: a batch study. J Environ Chem Eng 7(1):102904. https://doi.org/10.1016/j.jece.2019.102904
3. Abdelrady A, Bachwenkizi J, Sharma S, Sefelnasr A, Kennedy M (2020) The fate of heavy metals during bank filtration: effect of dissolved organic matter. J Water Process Eng 38:101563. https://doi.org/10.1016/j.jwpe.2020.101563
4. Handl S, Kutlucinar KG, Allabashi R, Troyer C, Mayr E, Langergraber G, Hann S, Perfler R (2023) Importance of hydraulic travel time for the evaluation of organic compounds removal in bank filtration. Chemosphere 137852. https://doi.org/10.1016/j.chemosphere.2020.127706
5. Hiscock KM, Grischek T (2002) Attenuation of groundwater pollution by bank filtration. J Hydrol 266(3–4):139–144
6. Jiang Q, Wang B, Jin G, Tang H, Xu J, Zhu X (2023) The response patterns of riverbank to the components carried by different pollution sources in the river: experiments and models. J Hydrol 617:128903. https://doi.org/10.1016/j.jhydrol.2022.128903
7. Stuyfzand P, Juhàsz-Holterman MA, De Lange W (2006) Riverbank filtration in the Netherlands: well fields, clogging and geochemical reactions. In: Hubbs S (ed) Riverbank filtration hydrology. Springer Netherlands, pp 119–153. https://doi.org/10.1007/978-1-4020-3938-6_6
8. Labelle L, Baudron P, Barbecot F, Bichai F, Masse-Dufresne J (2023) Identification of riverbank filtration sites at watershed scale: a geochemical and isotopic framework. Sci Total Environ 864:160964. https://doi.org/10.1016/j.scitotenv.2022.160964

9. Schelling M, Patil K, Boving TB (2023) Sustainable water treatment with induced bank filtration. Water 15(2):361. https://doi.org/10.3390/w15020361
10. Grischek T, Paufler S (2017) Prediction of iron release during riverbank filtration. J Hydrol Eng 9(5):317. https://doi.org/10.3390/w9050317
11. Szekely F, Nyiri G, Szucs P, Zakanyi B (2021) Analytically supported numerical modeling of horizontal and radial collector wells. J Hydrol Eng 26(12):04021040. https://doi.org/10.1061/(ASCE)HE.1943-5584.000213
12. Sharma B, Uniyal DP, Dobhal R, Kimothi PC, Grischek T (2010) A sustainable solution for safe drinking water through bank filtration technology in Uttarakhand, India. Curr Sci 107(7)
13. Gross-Wittke G, Gunkel A, Hoffmann A (2010) Temperature effects on bank filtration: redox conditions and physical-chemical parameters of pore water at Lake Tegel, Berlin, Germany. J Water Climate Change 01(1):55. https://doi.org/10.2166/wcc.2010.005
14. Grischek T, Schoenheinz D, Ray C (2003) Siting and design issues for riverbank filtration schemes. In: Hubbs S (ed) Riverbank filtration. Springer Netherlands, pp 291–302. https://doi.org/10.1007/0-306-48154-5_15
15. Srisuk K, Archwichai L, Pholkern K, Saraphirom P, Chusanatus S, Munyou S (2012) Groundwater resources development by riverbank filtration technology in Thailand. Int J Environ Rural Dev 3(1):155–161
16. Mohd F, Md Said MA, Yusoh R (2015) The development of a site suitability map for RBF location using remote sensing and GIS techniques. J Teknol (Sci Eng) 74(11)
17. Grischek T, Macheleidt W, Nestler W (2020) River bed specifics and their effect on bank filtration efficiency. In: Management of aquifer recharge for sustainability. CRC Press, pp 59–64
18. Bartak R, Grischek T, Ghodeif K, Wahaab R (2014) Shortcomings of the RBF Pilot Site in Dishna, Egypt. J Hydrol Eng 0(0):05014033. https://doi.org/10.1061/(ASCE)HE.1943-5584.00011
19. Poojitha SN, Hari Prasad KS, Ojha CSP (2022) Effect of clogging on riverbank filtration: an experimental analysis using Ganges riverbed sediment. J Hazard Toxic Radioact Waste 26(2):04021065. https://doi.org/10.1061/(ASCE)HZ.2153-5515.000068
20. Ray C, Melin G, Linsky RB (2002) Riverbank filtration: improving source-water quality. Dordrecht; Boston: Kluwer Academic Publishers; Fountain Valley, Calif.: In collaboration with NWRI, National Water Research Institute
21. Ray C, Shamrukh M (2011) Riverbank filtration for water security in desert countries. NATO Science for Peace and Security Series C: Environmental Security. https://doi.org/10.1007/978-94-007-0026-0
22. Oberleitner D, Stütz L, Schulz W, Bergmann A, Achten C (2020) Seasonal performance assessment of four riverbank filtration sites by combined non-target and effect-directed analysis. Chemosphere 261:127706. https://doi.org/10.1016/j.chemosphere.2020.127706
23. Schubert J (2003) Water-quality improvements with riverbank filtration at Düsseldorf waterworks in Germany. In: Ray C, Melin G, Linsky R (eds) Riverbank filtration. Springer Netherlands, pp 267–277. https://doi.org/10.1007/0-306-48154-5_13
24. Deyi M (2012) Development of guidelines for design and performance analysis of riverbank filtration for water treatment. UNESCO-IHE Institute for Water Education
25. Petts G (1996) Applied River morphology, edited by D. Rosgen. Pagosa Springs, CO: Woldland Hydology. Stream Reconnaissance Handbook, edited by C.R. Thorne. Wiley, Chichester, pp 373–374
26. Matusiak M, Dragon K, Gorski J, Kruc-Fijałkowska R, Przybylek J (2021) Surface water and groundwater interaction at long-term exploited riverbank filtration site based on groundwater flow modelling (Mosina-Krajkowo, Poland). J Hydrol Reg Stud 37:100882. https://doi.org/10.1016/j.ejrh.2021.100882
27. Zhang Y, Hubbard S, Finsterle S (2011) Factors governing sustainable groundwater pumping near a River. Ground Water 49(3):432–444
28. Caldwell T (2006) Presentation of data for factors significant to yield from several riverbank filtration systems in the U.S. and Europe. In: Hubbs S (ed) Riverbank filtration hydrology. Springer Netherlands, pp 299–344. https://doi.org/10.1007/978-1-4020-3938-6_14

29. Nagy-Kovács Z, Davidesz J, Czihat-Mártonné K, Till G, Fleit E, Grischek T (2019) Water quality changes during riverbank filtration in Budapest, Hungary. Water 11(2):302. https://doi.org/10.3390/w11020302
30. Schubert J (2002) Hydraulic aspects of riverbank filtration—Field studies. J Hydrol 266(3–4):145–161. https://doi.org/10.1016/S0022-1694(02)00159-2
31. Zhu Y, Zhai Y, Teng Y, Wang G, Du Q, Wang J, Yang G (2020) Water supply safety of riverbank filtration wells under the impact of surface water-groundwater interaction: evidence from long-term field pumping tests. Sci Total Environ 711:135141. https://doi.org/10.1016/j.scitotenv.2019.135141
32. Abdelrady A, Sharma S, Sefelnasr A, El-Rawy M, Kennedy M (2020) Analysis of the performance of bank filtration for water supply in arid climates: case study in Egypt. Water 12(6):1816. https://doi.org/10.3390/w12061816
33. Abdelrady A (2020) Effectiveness of bank filtration for water supply in arid climates: a case study in Egypt. CRC Press
34. Greskowiak J (2006) Reactive transport processes in artificially recharged aquifers (field and modelling studies). Humboldt-University, Berlin, Germany
35. Zuo R, Xue Z, Wang J, Meng L, Zhao X, Pan M, Cai W (2022) Spatiotemporal variations of redox conditions and microbial responses in a typical river bank filtration system with high Fe2+ and Mn2+ contents. J Hydrol 609:127777. https://doi.org/10.1016/j.jhydrol.2022.127777
36. Abdelrady A, Sharma S, Sefelnasr A, Kennedy M (2020) Characterisation of the impact of dissolved organic matter on iron, manganese, and arsenic mobilisation during bank filtration. J Environ Manag 258:110003. https://doi.org/10.1016/j.jenvman.2019.110003
37. Obeid AA, Abdelrady A, El-Rawy M (2022) Hydrochemistry and hydrogeology aspects of Alluvial Aquifer in Aswan City, Egypt. In: sustainability of groundwater in the Nile Valley, Egypt. Springer, pp 121–139. https://doi.org/10.1007/978-3-031-12676-5_6
38. Shamrukh M, Abdel-Wahab A (2011) Water pollution and riverbank filtration for water supply along river Nile, Egypt. In: Shamrukh M (ed) Riverbank filtration for water security in desert countries. Springer, Netherlands, pp 5–28
39. Ghodeif K, Grischek T, Bartak R, Wahaab R, Herlitzius J (2016) Potential of river bank filtration (RBF) in Egypt. Environ Earth Sci 75(8):1–13
40. Abdelrady A, Sharma S, Sefelnasr A, Kennedy M (2018) The fate of dissolved organic matter (DOM) during bank filtration under different environmental conditions: batch and column studies. Water 10(12):1730
41. Blanford W, Boving T, Al-Ghazawi Z, Shawaqfah M, Al-Rashdan J, Saadoun I, Schijven J, Ababneh Q (2010) River bank filtration for protection of Jordanian surface and groundwater. In: World Environmental and Water Resources Congress 2010, pp 776–781
42. Al-Jaf H (2022) Assessment of riverbank filtration for Sirwan River in Iraq. Int J Energy Water Resour 6(1):103–112. https://doi.org/10.1007/s42108-021-00149-x
43. Ghazal KA, Salman KA, Nieama AS (2023) Assessing the attenuation of microbial contaminants of Al-Kufa River water through the natural process of riverbank filtration. Egypt J Aquat Res 49(1):41–47
44. Abdel Wahaab R, Salah A, Grischek T (2019) Water quality changes during the initial operating phase of riverbank filtration sites in upper Egypt. Water 11(6):1258. https://doi.org/10.3390/w11061258
45. Bartak R, Grischek T, Ghodeif K, Ray C (2012) Beach sand filtration as pre-treatment for RO desalination. http://digitalcommons.unl.edu/civilengfacpub

The Status of the Applications of the Managed Aquifer Recharge in the Sultanate of Oman

Osama Ragab and Salma Al Maghawry

Abstract Water resources in arid regions have been an urgent issue over the last decades. Oman's geography and climate make it highly susceptible to water scarcity, prompting the need for sustainable water management practices. Water is the main requirement for all basic needs of any community including food production, industrial activities, and overall economic development. Water supply systems in Sultanate of Oman are subjected to several challenges that lead to the contamination and deterioration of water, such as agricultural activities, making the quality of available water resources not suitable for domestic use. Several factors affect these water resources, but the dry arid climates combined with the increasing issue of climate change and the rapid increase in population are the main factors since rainfalls are unpredictable and infrequent. The low levels of annual rainfall rates affect the level of groundwater aquifers since rainfall is their main source of recharge. This led to an interest in artificial groundwater recharge to prevent the constant decrease in the groundwater levels, prevent seawater intrusion, and enhance the water supply. The government's response to these issues has been through initiating desalination plants throughout the country, constructing wastewater treatment plants, and introducing the Managed Aquifer Recharge (MAR) and Aquifer Storage and Recovery (ASR). These techniques introduce the concept of water recycling, reduce land use and the risk of contamination, and increase the water supply that helps meet the increasing water demands. This chapter reviews conventional and non-conventional water resources and assesses the effectiveness of various water systems and future strategies in the water resources management sector in the Sultanate of Oman, including the use of ASR and MAR techniques.

O. Ragab (✉)
Hydraulic and Irrigation Structures Department, Civil Engineering, Faculty of Engineering, Alexandria University, Alexandria, Egypt
e-mail: osamaragab24@gmail.com

S. Al Maghawry
Civil Engineering Program, Faculty of Engineering, Sohar University, Sohar, Oman
e-mail: 181291@students.su.edu.om

M. El-Rawy and A. Negm (eds.), *Managed Aquifer Recharge in MENA Countries*, Earth and Environmental Sciences Library,
https://doi.org/10.1007/978-3-031-58764-1_6

Keywords Water resources · Groundwater · MAR · Aflaj · Dams · Sultanate of Oman

1 Introduction

Water is the main source of life in any nation [1]. Recently, the increasing water demand, rapid population expansion, and the limitation of available water resources have been urgent issues that have led to the obstruction of economic development and the civilization of societies [2]. Generally, in the Middle East and North Africa (MENA) region, analyses have shown that there are several problems facing the water resources and water systems [3, 4]. These issues represent an obstacle to the country's path of sustainable economic development [5]. Especially in arid and semi-arid regions—like the sultanate of Oman—there has been a valid concern regarding water shortage problems due to the long drought periods and the increasing water demand. In 2008, according to FAO [6], the long-term average of the total available water in Oman was found to be approximately 1,556 million cubic meters (MCM) per year, with 87% of the natural resources being due to precipitation. In comparison, the total water consumption in Oman is 1,873 MCM per year showing a noticeable shortage in water available and indicating the necessity of studying the available water resources in Oman and the water systems used to manage these resources, especially since the current increase in industrial and agricultural activities has led to the contamination and to the decrease in the quality of the available water resources making it rather hard to treat and provide water for the general domestic use [7].

Since Oman is an arid region, water resources are considered to be not adequate. Consequently, the development of various sectors, including industrial, agricultural, and irrigation, is threatened. Groundwater is known to be the main water resource in Oman. Hence groundwater is overused, especially under the current pressure of rapid population increase. As a result, the aquifer storages have been undergoing depletion, and this has caused seawater intrusion. Several solutions have been applied to solve the water insufficient issue, including the construction of recharge dams, and the desalination of seawater [8–11]. However, these solutions are expensive and mainly depend on the rainfall water. And since the rainfall in Oman is irregular and low, these solutions sometimes do not perform efficiently to solve the water issue in the country. In the light of the aforementioned, Managed Aquifer Recharge (MAR) has been introduced as a water management technique to increase the aquifer storage amounts. It is a technique that has been used in various countries to enhance the water supply in arid and semi-arid regions [9, 12–15]. MAR is the intentional artificial recharge of water into an aquifer by injection or infiltration. In this technique, a number of methods are used to artificially recharge the aquifer, but the conventional methods are surface infiltration and injection wells. The method of recharging aquifers using injection and recovery wells is known as Aquifer storage and recovery (ASR) and has been used to store water for seasons when the demand of water is higher than the available supply [15].

Several studies have reviewed the available water resources in various studies around the world. In Uganda, water resources are estimated at 66 km^3 per year with around 16% of the total land area of the country being open water bodies. Being a part of the Nile, Uganda has a network of rivers due to the high amounts of annual rainfall with an average of 1180 mm per year. Groundwater also represents a significant portion of water resources in Uganda, in which 61% of the water available in the country is in the form of springs and boreholes [16]. Another study in Ghana reviewed the water resources in terms of the availability, quality, and management of freshwater. The study stated that Ghana consists of three outlet systems, the Coastal River systems, Volta, and the South-Western. Moreover, the findings of the study showed higher microbial contamination of freshwater in surface water resources than in Groundwater in Ghana [7]. In Greece, water resources are more affected by the extreme weather conditions represented in floods, drought, and soil erosion. There is a noticeable shortage of water suitable for domestic use due to the poor quality of available water bodies [17]. Another study in Nigeria showed that the freshwater resources in the country are estimated to be around 286.2 km^3 per year. However, the main issue in the country is that resources are not evenly distributed resulting in only about 69% of the population having access to water [18]. In China, the water demand has been increasing due to rapid economic development. The main water resource issue in China was found to be in the North China plain, the efficiency of utilizing the water resources was found to be very low. It was recommended that more strict constraints on water resources be stated in terms of agricultural, industrial, and domestic activities [2]. Around the world, water resources are in excess of about 43,750 km^3 per year, distributed in all regions based on each country's physical features and climatic conditions [7]. In terms of freshwater resources, it was found that the USA has the largest portion of freshwater resources with around 45% meaning that freshwater resources are not evenly distributed around the world [19].

There have been several studies and research regarding the use of MAR and ASR around the world [9, 11–13, 20]. A review discussed the issue of groundwater in the MENA region where the demand on the water is way more than the available resources. The study investigates the challenges facing the water resources management in MENA region which are represented in the increasing population and the impacts of climate change. It shows that MAR have been a valuable solution through the recharge of dams, especially that more than 5% of dame in the Middle east are installed on main wadis for groundwater recharge purposes [12]. Another study has explored the MAR through checks on dames across Arani River in India. The effectiveness of the MAR in recharging the ground water table. The study showed a 1.5 m increase in the groundwater table proving that MAR has the potential for enhancing water availability and quality [11]. Another case study was made in Addis Ababa, Ethiopia, which is a city facing many challenges due to the rapid increase of population. This study employes a treatment plant in which the recharge is made using treated wastewater to improve the quality of the ground water. It was found that the use of the treated water showed promising results to reduce the total dissolved solids overall, improving the quality of water. The implementation of a treatment plant is an application of sustainability and can help maintain the water table [9].

Moreover, another study has reviewed the potential of MAR as a solution for stressed groundwater resources in California. MAR is expected to improve the water table levels, enhance the water quality, and increase the availability of water, especially in periods when the demand is relatively high [13]. In light of the aforementioned, it is very important to dedicate plans and budget towards overcoming the challenges regarding water resources, especially groundwater, to achieve a balance between the water demand by the public and the quantity of water available and conserving the natural groundwater aquifers [20]. This chapter aims to review the available water resources in Oman, assessing the mechanisms of current water systems, and the effectiveness of future strategies in the water management sector, including the potential of MAR and ASR techniques in Oman, especially using treated wastewater.

2 Study Area Description

Sultanate of Oman is located in the southeastern corner of the Arabian Peninsula in Asia and has a total land area of 309,500 km^2. It is surrounded by sea from two sides overlooking the Persian Gulf, the Gulf of Oman, and the Arabian Sea. It has a coast length of almost 3,165 km.

Topographically, Oman can be divided into three regions, the coastal region represented in Al Batinah Plain in the north and Salalah Plain in the South, with an elevation range of 0–500 m. Secondly comes the mountain regions, covering 15% of the country's total area and having a peak elevation of 3,000 m at the top of Al Jabal Al Akhdar. Finally, there are the internal regions with elevations not exceeding 500 m, and they represent the highest percentage of land (82%). This region is mostly covered with desert and sand and experiences very little precipitation yearly [21].

Oman is well known for the variation of climate from one region to another due to the large land area. It is mostly hot but humid in the coastal areas but hot and dry in the internal regions during summer. However, in Dhofar region in the South, the climate remains moderate all year round. Rain seasons in Oman are usually from October to April mostly occurring as tropical cyclones and flash floods develop in the Indian Ocean due to the relatively long coastal length of the country. However, Dhofar region experiences seasonal summer monsoons from June to September [22].

It is important to note that the country lacks any rivers passing through it and mainly depends on precipitation as the main source of water which is later stored as groundwater aquifers. Figure 1 shows the map of Oman including the main wadis passing through it.

3 Water Resources in Oman

In the following subsection, we will present and discuss the main sources of conventional water resources and non-conventional water resources in Oman.

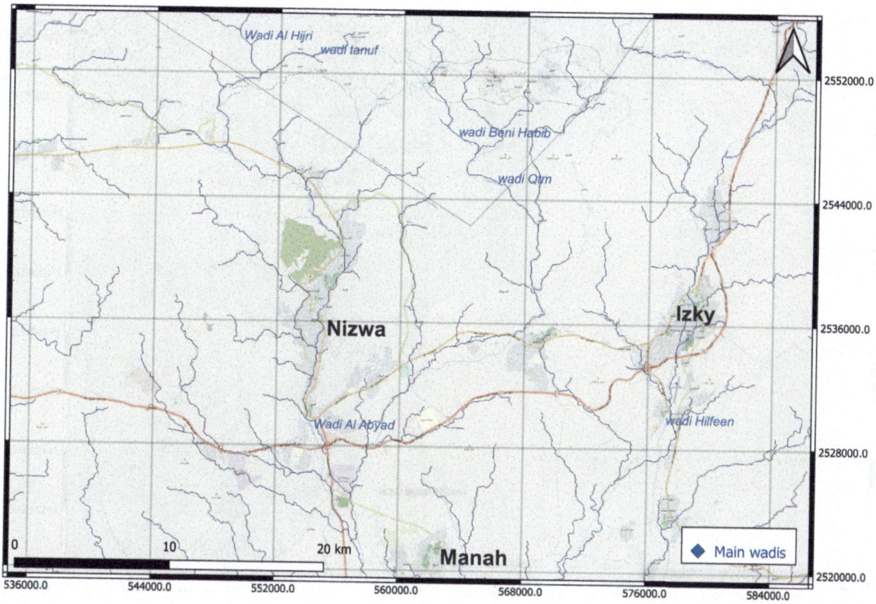

Fig. 1 Nizwa's hydrological map generated using QGIS software showing all main wadis and streams

3.1 Conventional Water Resources

3.1.1 Rainfall

Considering that Oman is an arid country with mostly dry weather, the rain amounts here are considered to be very low. The main sources of rain are escalating water storms, cold low pressures, monsoon wind currents, and equatorial cyclones. Average amounts of annual precipitation range from 50 to 100 mm with the exception of mountainous regions in which precipitation amounts reach almost 350 mm yearly [3]. This shows the effect topography has on the distribution of rain, in which the differences in amounts of rainfall in different regions are wide regardless of the distance between them.

In 2017, rain provided a total volume of 19,190 MCM per year. Almost 79% of it evaporated due to the high temperature, leaving just 3,288 MCM per year as effective rainfall, which usually generates surface water and groundwater aquifers due to of infiltration [3].

3.1.2 Surface Water

Surface water contributes to only 6% of conventional water resources in Oman. The volume of surface water flow is found to be 1050 MCM per year [23]. Surface water

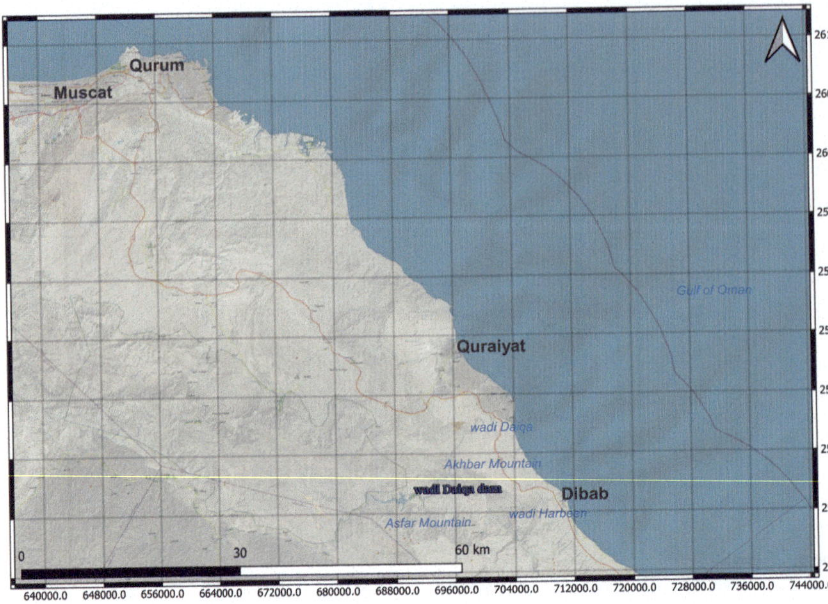

Fig. 2 Location of wadi Dhaiqa generated by QGIS

is represented by wadis and storage dams generated by floodwater. However, most floodwater leaks through the soil and is stored as groundwater.

Wadis are the main form of surface water in Oman. However, it is considered to be temporary because the flow in them usually occurs for several hours or days only after a flood event each year. The quantity of water flowing in wadis depends on the intensity and duration of the flood as well as the natural and geographic characteristics of the earth's surface. For instance, a layer of base rocks with low impermeability and a cliff with no plant cover allow a very suitable environment for surface flow. Hence why many wadis flow from mountains towards the sea or the desert.

Regardless of the temporary flow in wadis, these flows are still important because they feed natural and artificial wide catchment basins that exist beneath the ready lower parts. Significant wadis in Oman include wadi Dhaiqa that have an average flow of 60 MCM per year. In addition to Halfayan wadi with a catchment area of 4,373 km^2 [23] (Fig. 2).

3.1.3 Groundwater

Groundwater is water existing beneath the Earth's surface in the soil void spaces. It can be found in sedimentary surface spaces such as sand, gravel, coastal plains sedimentary rocks, and even in bedrock fractures. It is basically a rock formation that stores water. The amount of groundwater in Oman varies depending on precipitation amounts and the geology of the area. Most groundwater is hidden from view.

However, it can occasionally make an appearance in the form of springs [8] (Figs. 3 and 4).

In Oman, recharges to the natural aquifers are mainly represented in rain-water infiltrating through the ground due to the very high infiltration capacity of course alluvium and fissured rock despite the high percentage of evaporation due to the high temperature [8]. Other recharge methods include indirect recharge from

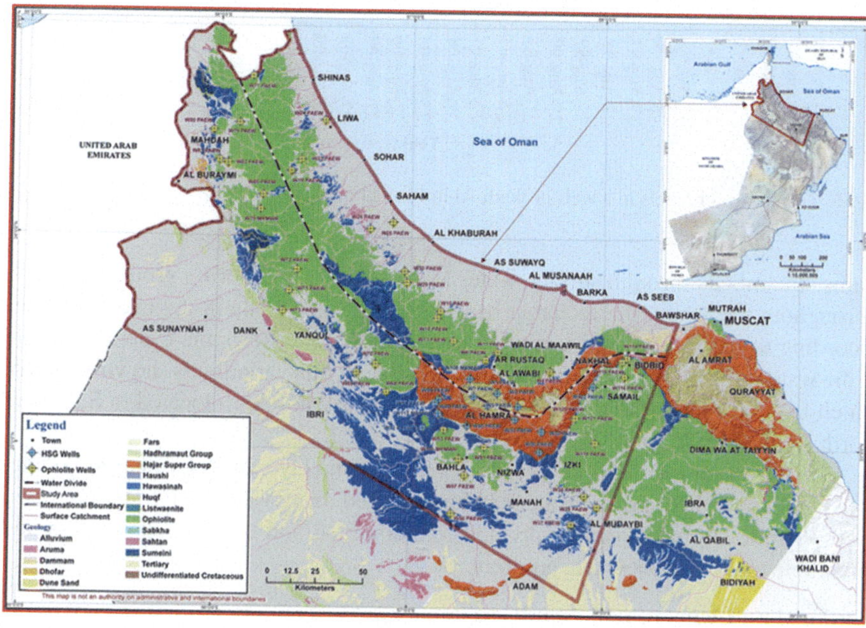

Fig. 3 Map showing the location of groundwater wells in north Al Batinah [8]

Fig. 4 Groundwater table including the saturated and unsaturated zones

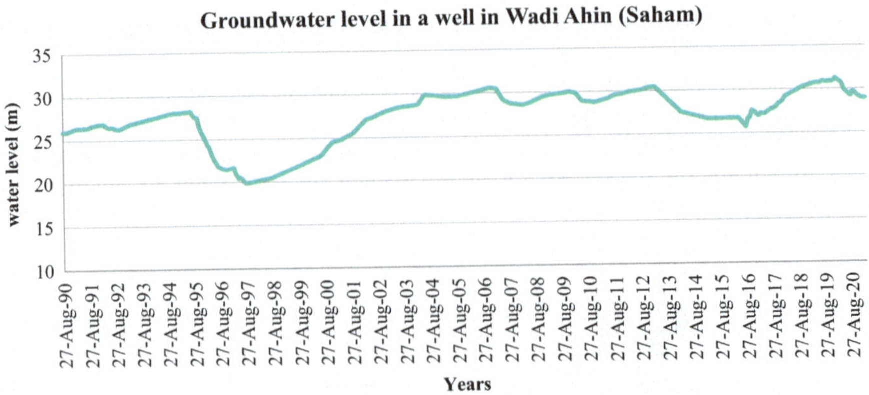

Fig. 5 Groundwater levels in a well in Wadi Ahin located in Saham

wadis streambeds through surface water leakage, and neighboring ground aquifers. Throughout the years, the recharge of the groundwater wells in the country varies according to the rainfall and temperature variation that affects the evaporation rate of the water infiltrated in these wells. An example of this variation is shown in Fig. 5, which indicates the groundwater levels in a well in Wadi Ahin located in Saham in North Al Batinah Region.

A groundwater aquifer is known as an underground layer of water-bearing permeable rock from which groundwater can be extracted using water well to be later used by the public. Aquifers can be divided into confined and unconfined aquifers and renewable and non-renewable aquifers.

Confined and Unconfined Aquifers

An unconfined aquifer is an aquifer that has the water table as an upper boundary in which the aquifer is under air pressure only. Consequently, the water level in a well is the same as outside the well. In contrast, a confined aquifer is an aquifer trapped between two impermeable soil layers meaning that water is under pressure that is higher than air pressure (Fig. 6).

Renewable and non-renewable Aquifers

Renewable aquifers are those fed by rainfall or surface water flow. However, non-renewable aquifers almost receive no recharge at all, which means that they only exist due to ancient times when Oman's climate was more humid than it is now. Renewable ground aquifers feeding water in Oman are found to be almost 1,300 MCM per year [8]. Their occurrence depends on the intensity of rainfall and the surface water flow. In areas where both of these parameters are high, there is a high probability of the existence of a new renewable aquifer if there is a permeable upper soil layer [3].

The main aquifer systems in Oman include the alluvial aquifers when water runs from the mountains to the coast, sand aquifer systems in the interior areas of Oman, and the underlying regional aquifers of the Fars complex and the al-Radhuma

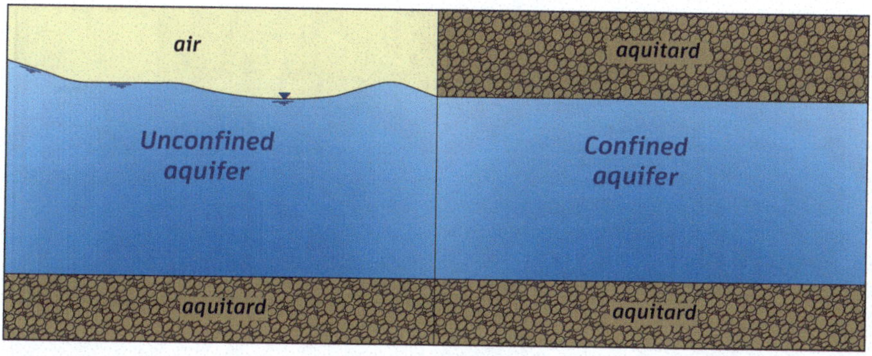

Fig. 6 Unconfined and confined aquifers

systems. Generally, the main aquifer water flows are formed in the Al Hajar Al Garbi Mountains in the north, and the Dhofar mountains in the South. In Both mountainous regions, the wadis' streambed consisting of sedimentary rocks represents a major factor in it is being a renewable water source due to the recharge the aquifer received from previously mentioned sources. The coastal plains also have ground aquifers lying beneath them that are fed by the flow toward the sea. In a similar way, the flow toward interior regions feeds shallow ground aquifers too [24].

Non-renewable resources in Oman are represented in carbon formations dating back to the third period whose thickness exceeds hundreds of meters. Water in these aquifers is mostly salty water with very low present recharge. These non-renewable aquifers exist in Dhofar (Najd), as shown in Fig. 7, Al Dahra (Al Massarat), and Al Shaqiya (Rimal Al Sharqiya). Al Massarat aquifer covers an area of 12,560 km^2 with a maximum depth of 400 m and two well fields that have been installed, Ain and Abu Karbah. Moreover, Al Shaqiya basin covers an area of 21546 km^2 with a very limited natural recharge of 10 MCM per year [24].

Challenges Facing the Aquifers in Oman

Groundwater is well known to be the main water resource used for all domestic, agricultural, and industrial purposes in Oman. For these uses, wells and well fields contribute almost 20% of the water supply and are managed by the public authority for electrical and water (PAEW). For instance, there are 744 wells supplying water to the distribution system in Oman [25].

As of recently, due to population expansion, there has been a noticeable increase in water demand, resulting in an increase in the rate of aquifer drainage via wells, causing a disruption in natural balance. When this occurs in coastal areas, it leads to the mixture of seawater with aquifer water. Additionally, when the rate of groundwater flow is low, the passage of groundwater from high to low levels can take up to 100 years; during this period, the water is in constant direct contact with rocks which may lead to chemical processes resulting in water quality deterioration [6].

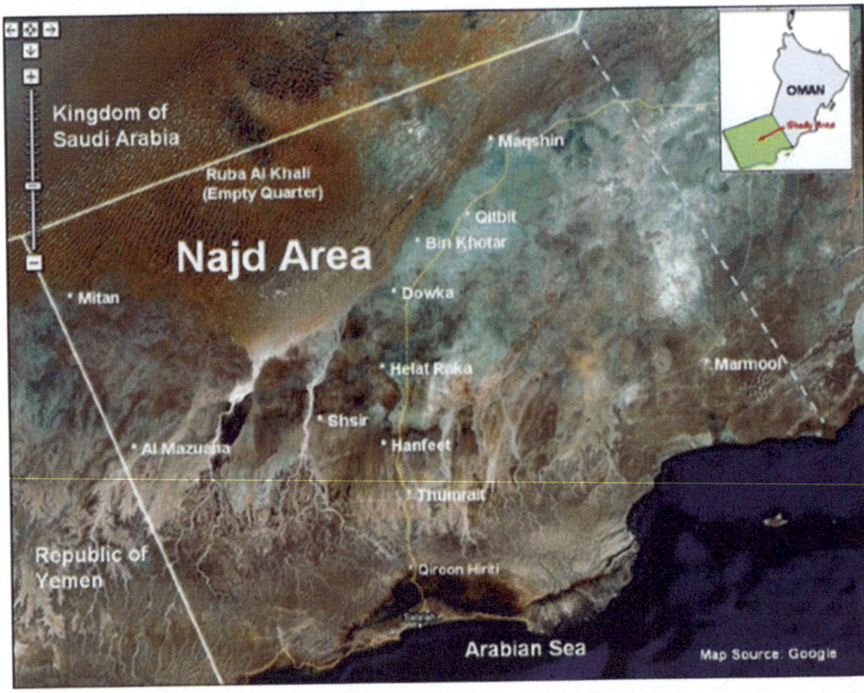

Fig. 7 Satellite image of Najd area [24]

3.1.4 Springs

Springs are groundwater discharged at the surface of the ground naturally, usually when water is under higher pressure than air pressure. Water springs in Oman have two geological systems, limes lime rocks and Ophiolite rocks in the Al Hajar Mountains. Lime rocks have very high porosity and cracks, which help water flow and ensure good quality. Ophiolites are less permeable and may yield alkaline water, as the case in Al Dakhiliya, Al Sharqiyah, and Al Batinah [4].

In Oman, springs are various depending on the water quality discharge and the temperature. In cases where the springs feed on groundwater in very deep layers, the water is heated before it surfaces. This occurs in springs that exist in the northern part of the Omani mountains. Springs play an important role in local life because they are used widely for drinking and irrigation purposes. In addition, Hot Springs are famous for treating diseases, and so it has been advertised as a tourist attraction. Overall, springs in Oman have a pH of 7.4–11.9 depending on their geological nature [3].

Famous Hot Springs include Ain Al Khasfa as shown in Fig. 8, in Al Rustaq, and Ain Al Thawarah. Ain Al Khasfa is known for its location while Ain Al Thawarah is known for its beautiful scenery and landscape. Both Hot Springs have been used for water supply purposes and therapeutic purposes due to the high quality of the water.

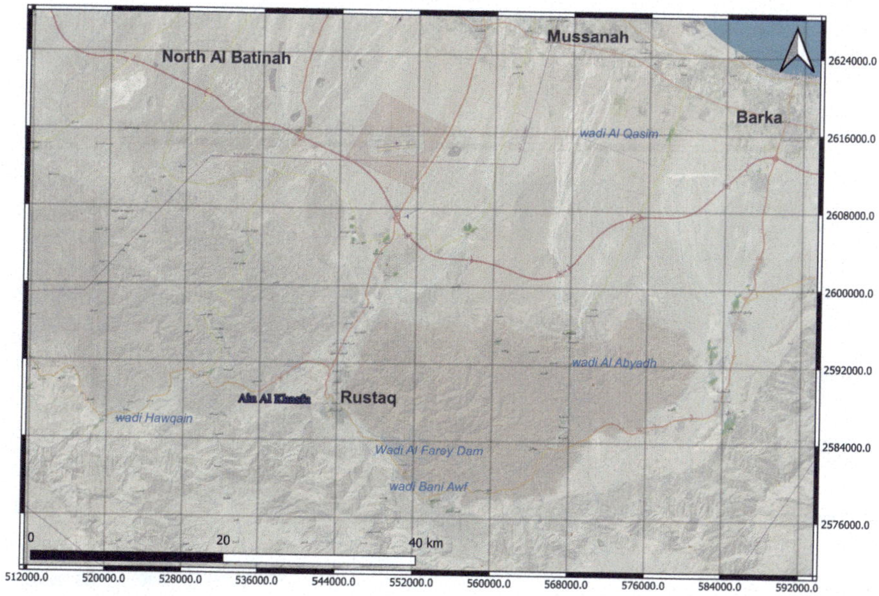

Fig. 8 Location of Ain Al Khasfa generated by QGIS

Cold springs in Oman include the spring of Ain Razat, also known for its attractive landscape and used for crop irrigation [3].

3.2 *Non-conventional Resources*

3.2.1 Desalinated Water

Desalination is the process of removing salt from seawater to make it suitable to use by humans for irrigation and industrial purposes. In regions with low conventional water resources, it is considered more economical to desalinate seawater and groundwater instead [23]. Although it may be considered a reliable option, there are still many challenges facing it, including the high cost and energy requirements and the requirement of large storage facilities, making this non-conventional resource only suitable in locations where it is impossible to get natural water access.

The first desalination plant was established in 1976 with a production capacity of 1.0 MCM per year. Since then, other desalination plants have also been established in coastal and interior regions with capacities ranging from 120 to 7200 m^3/day [25]. Specifically, a number of these salination units have been installed at private wells in Al Batinah. Over 90% of the total desalination capacity is installed in Al Gubrah reaching 52 MCM per year. Statically, the number of large desalination plants in

2010 was five and it increased to ten in 2019, As for the small desalination plants it increased from 33 to 49 in 2019 [25] (Fig. 9).

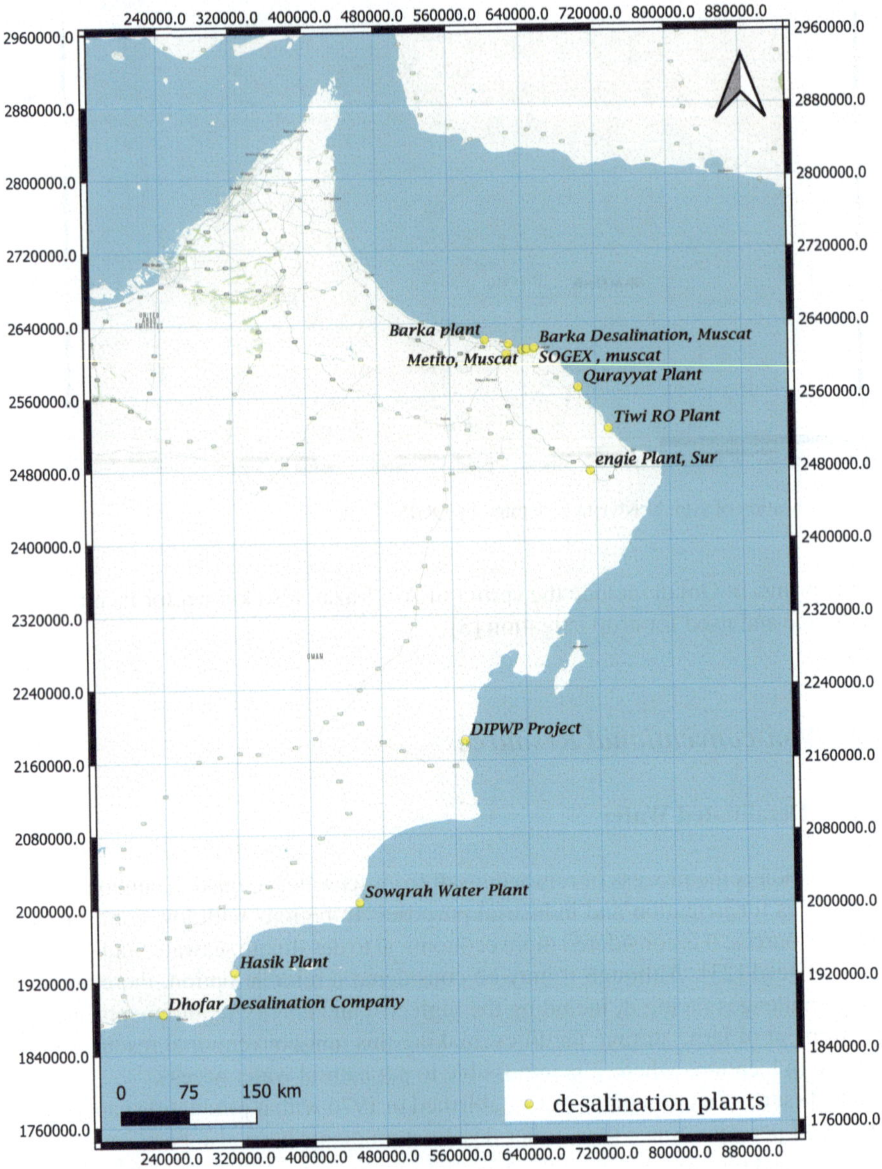

Fig. 9 Location of coastal desalination plants in Oman [26]

3.2.2 Treated Wastewater

In Oman, oil companies produce huge amounts of wastewater that need to be treated. This treated wastewater gives a great opportunity to increase water supply resources and decrease the groundwater drainage rate. Haya Water is a wastewater treatment company operating 58 sewage treatment plants all over the sultanate producing almost 200 MCM per year. There is a current plan to cover the entire Muscat region and to reach the production of 100 MCM by 2020 [27]. Additionally, more plants have been installed in Khasab, Ibri, Buraimi, and other regions in order to use the treated water for industrial and irrigation purposes.

4 Water Systems in Oman

Water systems are structures used to extract, distribute, and store water. These water management systems may be ancient or more advanced and sophisticated and they may serve a single purpose or be designed to serve many purposes regarding the watershed.

4.1 Al Zajirah

Al Zajirah was the main water extraction method from open wells until the invention of pumps in the 50s of the past centuries. And it is dependent on the utilization of the strength of animals to lift the water. Its main structure consists of two major (well wheels) made from pieces of wood and connected together with rope. The drawback of Al Zajirah is that it needs mechanical force to lift the water up under the supervision of the farmer.

4.2 Al Falaj

A channel system is dug into the ground to transport water via gravitational flow from one place to another. Historically, Afalajs have existed in Oman for over 1500 years some Afalajs even date back to 2500 years and have been the only way people were able to get a constant water supply. Due to their historical significance, Afalajs have been listed in the World Heritage list not just due to the falaj's channel itself but also the significance of the surroundings and their location. Included Afalajs are Falaj Daris and Falaj Al-Khatmeen in Willayat Nizwa Falaj Al-Malaki in Willayat Izki, Falaj Al-Mayssar in Willayat Al-Rustaq and Falaj Al-Jeela in Willayat Sur [25].

Afalajs comprises of usually three systems, the source which is usually a wadi, the transportation system, including several branches, and the delivery system. The main

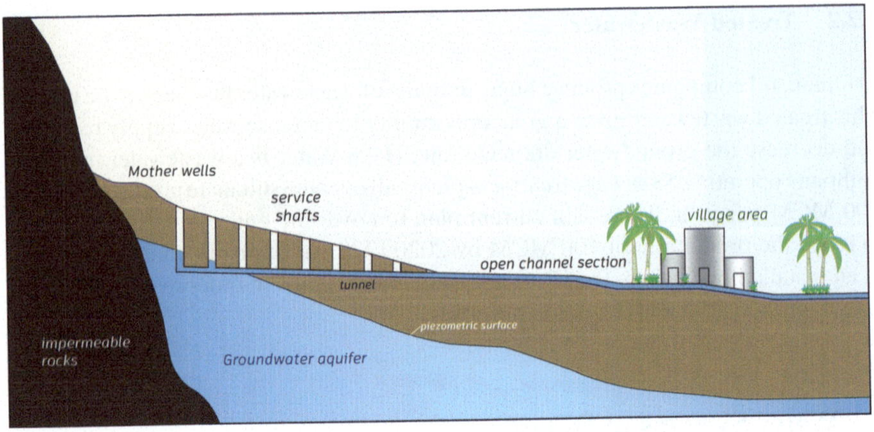

Fig. 10 Structure of Al Dawoodi falaj

channel, which is the source, usually exists in the mountains and carries water to the branches and subbranches connected to it. These main channels take into consideration minimizing the construction cost and avoiding adding any means to carry water and instead, transporting water naturally using the natural slope of elevated areas. The size and length of Afalajs depend on the number of uses it serves, usually, small Afalajs serve a couple of families only while large Afalajs may serve an entire village. Afalajs fed by massive groundwater aquifers do not depend on the temperature or the weather. Instead, once it starts running it rarely dries [25].

There are three main types of Afalajs in Oman. The first is Al-Dawoodi, built as an underground tunnel with several vertical outlets from the main channel to the targeted areas [25] (Fig. 10)

The second is Al Ghayli falaj which obtains water from the bottom of the wadi and is constructed as an open channel transforming water to the distribution section. This type depends on rainfall, so the flow may stop temporarily in dry seasons (Fig. 11).

Third is Al Ayni falaj, which is spring-fed, especially springs that come from limestones due to their high-quality water. Examples include A-Kastah spring in the wilayat of Rustaq and Al-Thawarah in the wilayat of Nakhl (Fig. 12).

There are 3000 Afalajs located in Oman with a supply of 552 MCM per year with some losses due to water leakage. Table 1 shows the number of operating and dry Falajs in Oman, while Table 2 shows a comparison between some famous falajs in Oman. Most Afalajs are located in the Al Hajar Mountains due to the topography of the area.

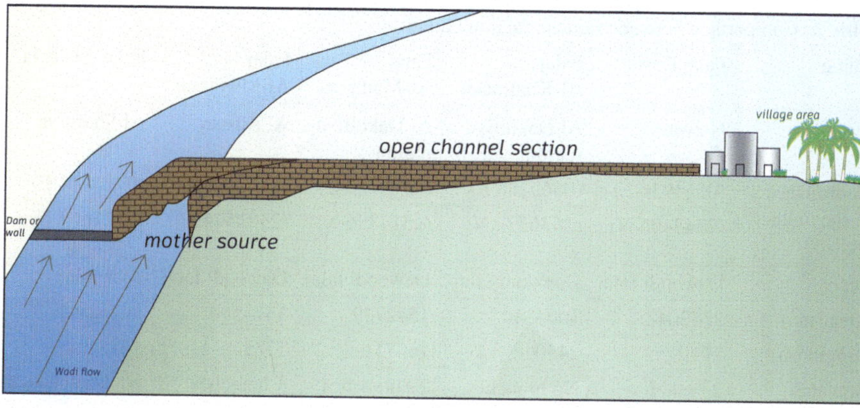

Fig. 11 Structure of Al Ghayli falaj

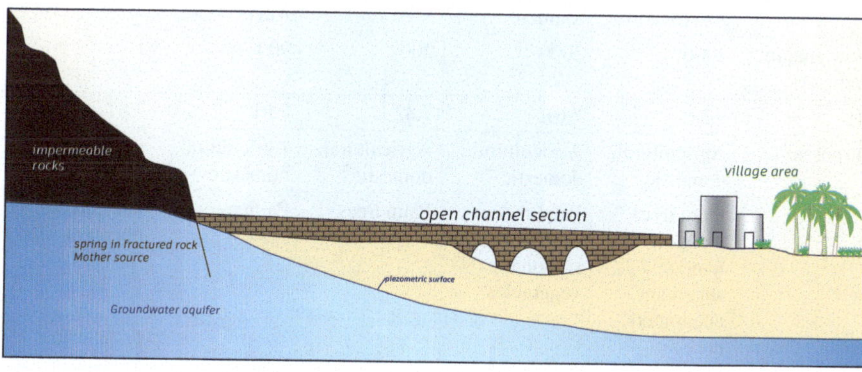

Fig. 12 Structure of Al Ayni falaj

Table 1 Numbers of operating and dry falajs [25]

Type	Number operating	Number dry	Total	Operating percentage (%)
Dawoodi	627	340	967	65
Aini	989	163	1152	85
Ghaily	1401	592	1993	70
Total	3017	1095	4112	74

4.3 Dams

4.3.1 Recharge Dams

Due to the higher temperatures in Oman, evaporation rates exceed rainfall rates. Hence, storing water under the ground is a more suitable idea than keeping it as

Table 2 Comparison between famous falaj in Oman

Name	Falaj Daris	Falaj Al-Khatmeen	Falaj Al-Malki	Falaj Al-Muyassar	Falaj Al Jeela
Place	Nizwa	Al Dakhiliya region	Al Dakhiliya region	Al Rustaq	Al Sharqiya
Coordinates of the mother well	(0554936 E) (2543496 N)	(0569288 E) (2536777 N)	(0579736 E) (2514896 N)	(0547631 E) (2582594 N)	(0723172 E) (2521664 N)
Type	Dawoodi falaj	Dawoodi falaj	Dawoodi falaj	Dawoodi falaj	Aini falaj
Area (m^2)	2382642	1004340	1572739	1574580	309529.5
Length (m)	7990	2450	14875	5783	161
Branches	2 branches	No branches	17 branches	5 branches	No branches
Temperature (°C)	29	30	37	31.9	29
Source	Wadi Al-Abyadh	Wadi Al Maiden	–	Wadi Bani Auf	Wadi Shab
Flow amount (L/s)	2000	2000	200	900	1
pH	7.3	7.61	7.82	7.61	7.87
Purposes	Agricultural, domestic	Agricultural, domestic	Agricultural, domestic	Agricultural, domestic	Agricultural, domestic
Crop	Palm trees, mango, lemon, sugarcane, cucumbers, etc.	Palm trees, mango, lemon, seasonal vegetables	Palm trees	Palm trees	Palm trees, pomegranates

surface water. This Operation is named artificial ground feeding in which wadis' surface water is leaked under the surface of an alluvial channel. The storage capacity of recharging them exceeds the storage capacity of most surface establishments at a much lower cost. Examples in Oman include the Al Khawd dam in Al Seeb with a capacity of 11.6 MCM and Ma'awil in Barka with a capacity of 10 MCM [6].

4.3.2 Surface Storage Dams

Although evaporation rates in Oman are high, constructing surface storage dams in wadis that are constantly flowing—like wadi Dhaiqa—may be beneficial. As a result, almost 58 storage dams were constructed, with a storage size of almost 750,000 m^3. Examples include Al Shurayjah Dam, with a capacity of 10200 m^3, and Al Ayn, with a capacity of 3,600 m^3 [6].

4.3.3 Flood Protection Dams

Floods are a direct extreme danger towards lives and agriculture. Especially in Oman, with a rich history of flash floods due to high mountains, lack of plant cover, and very steep slopes which give water higher strength. Hence, several flood protection dams have been constructed in Oman. However, these dams must be designed to withstand very strong pressures, and each catchment must be studied carefully to determine the expected peak flow rate. Overall, there are a total of 184 dams in Oman. In 2020, the total dams' capacity reached 324 MCM. It is worth mentioning that wadi Dhaiqa alone has a capacity of 100 MCM [6].

5 Case Studies on MAR in Oman

Managed Aquifer Recharge (MAR) is a crucial sustainable water resource management technique, especially in arid regions like Oman, where water scarcity is a significant challenge. MAR involves the intentional recharge of aquifers with surface water via dams and/or treated wastewater to store excess water during wet periods and recover it during dry periods [14].

5.1 Dams and Reservoirs

Oman has been utilizing dams and reservoirs as storage structures. These water storage facilities capture rainfall and surface runoff, which is subsequently used for MAR and agricultural purposes. It not only caters to the local water supply but also recharges aquifers by allowing excess water to percolate into the subsurface. The two most important storage dams in Oman (Wadi al Khoud and Wadi Hilti/Salahi) are described further below.

5.1.1 Wadi Al Khoud Dam

Wadi al Khoud is located in Seeb in the Western region of Muscat at coordinates 23.629076 Latitudes and 58.163690 longitudes (Fig. 13). The area of the wadi is known for its dry climate with an average temperature range of 21–34 °C. Summers are known to be extremely hot with temperatures reaching almost degrees Celsius [28]. Another climatic feature of the area is the noticeably low annual precipitation amount of almost 100mm. Rainfall usually occurs in the months of winter (December to March) in the form of storm events that are characterized by their short period and high intensity. The soil in the area of the Wadi is mostly from the stones of the Hajar Mountains resulting in high consistencies of Ophiolites and limestones. The watershed of the Wadi has an area of around 1635 km^2.

Fig. 13 Wadi Al-Khoud location

Due to its length, Wadi Al Khoud is known as Wadi Samael at the source, Wadi Fanja at the middle, and Wadi Al Khoud at the end of it. Wadi Al Khoud dam is known to be the first pilot project in the recharge dams program in Sultanate of Oman. Due to the existing conditions at the time, several studies regarding water resources in the region have recommended constructing this recharge dam at Wadi Al Khoud. Consequently, the dam was built in a total of 15 months from December 1983 to March 1985 [28].

The dam is an earth dam with a compacted clay core that has a length of 5,100 m, a height of 11m, and a volume of around 1.0 MCM. The spillway of the dam is made of gabions with a total length of almost 3,000 m and a peak flow rate of 12500 m^3/s. Moreover, the reservoir of the dam has a huge storage capacity of 11.6 MCM with an area of around 3.2 km^2; the dam is constructed with weep holes each with a diameter of 1.2 m. These drainage holes are semi-closed and poured into a large basin in the direction parallel to the dam to ensure an even distribution of water behind the dam in order to recharge the groundwater aquifers in the area. The structure of the dam is shown in Fig. 14.

The dam was originally constructed as a recharge dam with no intention of using it as a flood protection dam. However, the dam's ability to store around 11.6 m^3 of water helped reduce the high runoff of the wadi and provided protection to the main streets of Al Seeb city against the frequent flash floods. Since the operation of the dam, there have been several developments that have resulted from the experience of the operation of the dam for more than 30 years now. For instance, iron blocks have been installed to control the drainage of water through the drainage holes as well

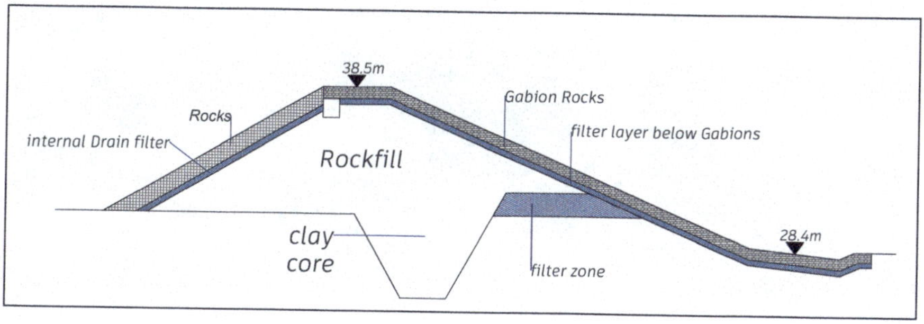

Fig. 14 Structure of Wadi Al Khoud dam

as increase the height of the reservoir of the dam to guarantee the even distribution of the flow. Moreover, in 1986, a monitoring device was installed to monitor the performance of the dam since it was originally initiated as a pilot project. Data collected by this monitoring device during the flood events in 2003 have shown that around 11.6 MCM of water was stored by the dam, if not, these massive amounts of water were going to end up in the sea with no way to take advantage of it to raise the levels of the groundwater aquifers. The noticeable success of this dam and its significant impact on the water resource in the area has encouraged the government to proceed with the construction of more recharge dams in the recharge dams' program (Fig. 15).

Fig. 15 Aerial photograph of wadi Al Khoud dam [9]

Fig. 16 Wadi Hilti/Salahi

5.1.2 Wadi Hilti/Salahi Dam

Wadi Hilti/Salahi is in the southern region of Sohar in north al Batinah at coordinates 24.279021 latitudes and 56.588232 longitudes (Fig. 16), North al Batinah is known to be an arid region with very limited water resources in which groundwater is the main source of water in the area. The area has a very humid hot climate during summer with very high temperatures that pass 40 °C. In the winter, the weather is slightly better with temperatures ranging from 15 to 25 °C [29]. The annual average amount of rainfall is 126 mm and -like Muscat-mainly comes from the tropical cyclones and the flash floods that strike the area. There is a highly variable terrain in the area due to the presence of Hajar Mountains [29]. The watershed of the wadi has an area of around 556 km^2.

Wadi Hilti/Salahi dam is the second recharge dam built in Sultanate of Oman. Which it was built to store the water caused by precipitation in flood season. The dam is constructed on two main wadis, Al Hilti, and AL Salahi, as well as secondary streams in North al Batinah. The dam was designed to recharge the area's groundwater aquifers and improve the water quality. The aquifers in north al Batinah have several wells that have been drilled for irrigation purposes. The Hilti /Salahi dam has been found to serve around 630 wells in the area. The construction of the dam took around nine months from November 1984 to August 1985 [29].

The dam has an effective capacity of 0.55 MCM, a length of around 9 km, a height of 4.5 m, and a total volume of 260,000 m^3 [25]. The spillway of the dam is constructed from gabions with a length of 1,120 m and a peak flow rate reaching 2,200 m^3/s. Furthermore, the reservoir of the dam has an area of 0.15 km^2 with 31 weep holes that drain water [23]. The dam has two water distribution areas one in Wadi Al-Hilti and the other in Wadi Al-Salahi. The structure of the dam itself was constructed from gabions and rockfill, in cases of floods, the drainage holes distribute water in a way that helps recharge the groundwater aquifers. In cases of overflooding, the dam is designed in a way that discharges the excess water to the sea so that it protects the residual areas surrounding the wadi. The structure of the dam is shown in Fig. 17.

Wadi Hilti/Salahi dam has been provided with a monitoring network that measures precipitation amounts in the upper catchment of the wadi and a device that measures

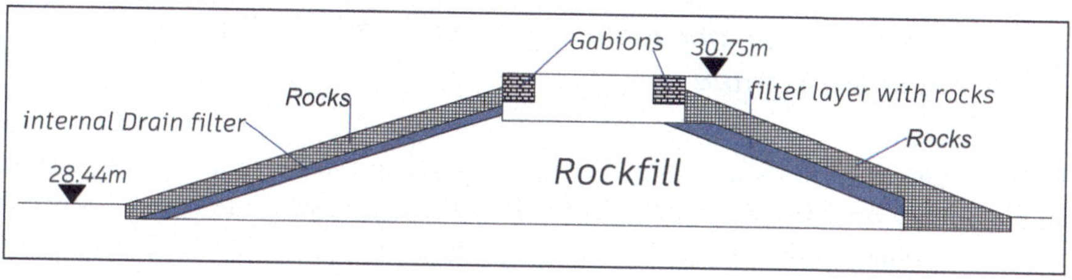

Fig. 17 Structure of Wadi Hilti/Salahi dam

the flow levels near the dam structure and the distribution areas. In addition, devices that measure the groundwater level have been installed too in order to evaluate the performance of the recharge dam. When comparing the results obtained from these monitoring devices to the feasibility study of the dam, it can be observed that the dam has shown an unexpectedly good performance in terms of the sediment accumulation in the lake of the dam as well as the recorded groundwater aquifers levels. The location of Wadi Hilti/Salahi is shown in Fig. 18.

Fig. 18 Location of Wadi Hilti/Salahi

5.2 Aquifer Storage and Recovery (ASR) and MAR Using Treated Wastewater

In ASR technique, wells are used for injecting surplus water during wet periods and extracting it during dry spells. It involves the injection of treated wastewater into underground wells during the rainy season, ensuring a sustainable source of water for agricultural and landscape irrigation during the dry season ASR systems sometimes use treated water as a common source [14]. The wastewater undergoes a treatment process in which contaminates are removed to minimize the pollution of ground water making it suitable to be stored in the aquifers until it is needed. When the water demand is high, the stored water is pumped back to the surface and is used for various purposes and uses. However, the treated water that is to be injected into aquifers in MAR and ASR techniques has to be within the established water quality standards and must comply with the local and national regulations in order to prevent the contamination of groundwater. Such practice increases the overall quality of the groundwater. It dilutes the salinity of the ground water and reduces the concentrations of unwanted matter in the water. Moreover, this is an application of sustainability in water resources management by reusing the treated water and storing it for use during periods when a shortage of water is experienced. MAR and ASR are approaches that promote the overall goal of sustainability and efficient water resource management.

Several research have been conducted in Oman studying the use of treated wastewater (TWW) in MAR and ASR applications. One study investigates the economic feasibility of using treated wastewater for MAR to attempt to solve the water scarcity issue in Muscat. The research studies the cost of injecting 85000 m^3 of TWW into aquifers daily. Main findings show that the estimated cost ranges from 0.353 to 0.550 dollars per cubic meter, showing the project's economic feasibility. The estimated cost is found to be lower compared to similar international projects. However, future recommendations were made to use further advanced treatment methods for TWW and to investigate the cost of applying this project for industrial use as well [30].

Further researches [32] include studies using MODFLOW; one study was made in the Jamma aquifer which is a costal aquifer in Oman to assess the feasibility of utilizing TWW in MAR to reduce the salinity of the Jamma aquifer. The Jamma aquifer was found to be susceptible to high salinity levels due to seawater infiltration into the aquifer and degradation of water quality, consequently affecting the irrigation and agricultural applications. MODFLOW was used in the study using simulated scenarios to find its effect on the aquifer. It was found that without any management the Jamma aquifer's state will become progressively worse in the upcoming 20 years. Using MAR with TWW was found to be hydrologically feasible to prevent the increasing salinity of the aquifer [31]. Moreover, another research using MODFLOW was made in Al Khawd costal area to evaluate the impact of MAR with TWW to help enhance the water supply and minimize the damage to aquifers. Numeric solutions and economic studies were made involving irrigation wells and injecting TWW. Findings show that by implementing MAR, the amounts of water for drinking purposes could double, and the possibility of using TWW with MAR is hydrologically and economically feasible [32].

Another case study was made in Samil lower catchment to maximize the recharge of the aquifer. This study used a simulation-optimization (S-O) modelling approach to address groundwater management problems. This study also shows the feasibility of MAR with TWW in Samil lower catchment. It shows that this technique could enhance water quality and availability for the domestic and agricultural use [33]. Additionally, a study was made to explore ASR and MAR's hydrological and economic feasibility using TWW in Muscat city, especially the stressed coastal areas where aquifers suffer from high salinity. Numerical solutions using MODFLOW were employed, and it was found that MAR using TWW can approximately double the abstracted volume of water for drinking purposes when managing agricultural well pumping and relocating domestic supply wells. The study showed that MAR represents a smart water management option [14].

Overall, MAR and ASR using TWW are valuable strategies and techniques in the water resources management sector, especially in arid regions like Oman. It has been proven in many case studies in Oman to be hydrologically and economically feasible and to have a very high potential to mitigate aquifer damage, maximize aquifer recharge and improve overall water quality.

6 Conclusions

In a country with such an arid climate like Oman, the freshwater supply mainly depends on rainfall stored in groundwater aquifers. Even precipitation is not frequent and mainly appears in the form of storms, tropical cyclones, and flash floods that mainly struck the northern coast and Salalah in the southern region of Oman. The climate conditions and growing population drive the main challenges of water resources in Oman.

Regarding Conventional water resources, it was found that Oman does not have any rivers passing through it and surface water is limited in the form of wadis. The amount of 1050 MCM was found to be the latest recorded value of the total volume of surface water flow in Sultanate of Oman. Although the flow of wadis is particularly dependable on precipitation and is considered temporary flow, this temporary flow is of great significance because it feeds and recharges the aquifers and catchments. In terms of non-conventional water resources, records show that the number of desalination plants has increased from 33 in 2010 to 49 in 2019. Furthermore, the current status of sewage treatment plants are fifty-eight plants with an annual production rate of 2000 MCM.

As for water systems in Oman, it was found that they existed in the form of Al Zajirah in the past and currently exist in the form of dams and Aflaj. Regarding dams. Oman has around 184 dams with a total capacity of 324 MCM in which wadi Dhaiqa dam alone has a capacity of 100 MCM. Moreover, 60% of the freshwater supply is provided by aflaj which makes them extremely important for the agricultural industry.

To address the challenges of water resource management in Oman, Oman has implemented various water management strategies, including the construction of Afalajs and Dams, and the introduction of managed aquifer recharge (MAR) and aquifer storage and recovery (ASR). The use of MAR and ASR with the utilization of treated waste tawer has been explored to recharge aquifers and improve the overall water quality and to provide more water supply, especially in coastal area where the aquifers suffer from salinity. Overall, studies have shown that the government is aiming towards initiating more water recycling projects that introduce the reclamation of wastewater. And MAR has gained prominence as a viable solution for replenishing and managing groundwater resources in Oman.

7 Recommendations

To preserve the current waste resources and minimize the water loss in sultanate of Oman, there is an urgent need to develop effective water management systems. Desalination plant production should be expanded and used to obtain fresh drinking water due to the fact that the process of desalinating water is quite costly. Hence freshwater of high quality should be only used for drinking and human purposes. Furthermore, additional wastewater treatments should be installed with higher capacities and production rates as well as constant monitoring systems that ensure that the water produced is of sufficient quality and quantity. The irrigation systems need to be more developed, and the traditional ways of harvesting should be replaced with modern advanced rainwater harvesting systems that improve the use of water and minimize the wasted amounts.

The groundwater aquifers are currently only fed by rainfall water that cannot be controlled. Hence, artificial recharge of these aquifers by surface water should be adopted in order to preserve the high levels of the groundwater which the country depends on as the main source of drinking water. Additionally, considering that dams are of significant importance to the country, regular monitoring programs and restoration should be conducted in addition to the construction of retention and detention dams in the active wadis of the main catchments [4]. Aflaj should also undergo regular maintenance as they are considered a main water system and a significant tourist attraction of the county. Studies have needed to that new techniques have been introduced in the irrigation engineering aspect. For instance, artificial pools and consolidated farming could be utilized by aflaj water to help meet the water demands in the agricultural sector [34].

Furthermore, mandatory laws must be employed to manage the country's water resources. in which customary laws in the commercial, industrial, and education systems must be made to adopt sustainable water practices. Policymakers must adopt measures to increase water use efficiency, develop an approach to identify the significant issues and set a list of possible solutions to overcome the obstacles in the path of an effective water resource management system.

Additionally, given the success shown in several case studies, it is recommended to explore the implementation of MAR and ASR techniques further, which have proven to be hydrologically and economically feasible. These MAR and ASR projects should also include the use of treated wastewater to improve the quality of the ground water aquifers. However, further investigations into advanced treatment options are needed to ensure that the effluent meets the regulations and standards. To ensure development, there should be integrated water resource management involving the use of recharging dams, MAR, and ASR, along with sustainable groundwater use, to ensure a comprehensive development. by implementing these enhance techniques, mitigating the challenges and encouraging sustainability, Oman can move towards a more resilient water future.

Moreover, social awareness regarding the water situation in Oman should be increased through involvement in the management process, social media campaigns, coordination between the government and the educational institutions, strengthening community participation through conferences and training programs as well as the organization between the government and the private sector and private companies [5]. Hydrological knowledge should be paid greater attention to in which investments in the water resources aspect should be mandatory for the private sector. Further research regarding the possible further water management technique available in Oman and their effectiveness as well as the performance of the hydraulic systems used to evaluate the practical solutions that help provide higher efficiency and minimize the effects of the current climate change. The protection of the available water resources must be set as a high priority to ensure sustainable economic development and to guarantee the quantity and quality of water variables for the current and future generations.

References

1. Allon F (2020) Water resources. In: International encyclopedia of human geography, 2nd ed, pp 233–240
2. Li Y, Han Y, Liu B, Li H, Du X, Wang Q, Wang X, Zhu X (2023) Construction and application of a refined model for the optimal allocation of water resources—Taking Guantao County, China as an example. Ecol Indic 146:109929. https://doi.org/10.1016/j.ecolind.2023.109929
3. Abushammala MM, Al-Bulushi MM, Qazi WA (2019) Water loss management in Muscat: case study of Al-Seeb water supply system. In: Oman in the 21st century: issues and challenges. Nova Science Publishers, Inc., pp 229–277
4. Al-Kalbani M, Saif M, Price MF, Timothy O, Mushtaque A, Asma A (2016) Integrated environmental assessment to explore water resources management in Al Jabal Al Akhdar, Sultanate of Oman. Reg Environ Change 16:1345–1361. https://doi.org/10.1007/s10113-015-0864-4
5. Murad AA, Nuaimi HA, Hammadi MA (2007) Comprehensive assessment of water resources in the United Arab Emirates (UAE). Water Resour Manag 21:1449–1463. https://doi.org/10.1007/s11269-006-9093-4
6. FAO (2021) AQUASTAT Country profile—Oman. FAO, Rome. https://www.fao.org/3/cb4413en/cb4413en.pdf

7. Yeleliere E, Cobbina SJ, Duwiejuah AB (2018) Review of Ghana's water resources: the quality, no, and management with particular focus on freshwater resources. Appl Water Sci 8:93. https://doi.org/10.1007/s13201-018-0736-4
8. Abdalla O, Abri RA, Semhi K, Hosni TA, Amerjeed M, Clark I (2016) Groundwater recharge to ophiolite aquifer in North Oman: constrained by stable isotopes and geochemistry. Environ Earth Sci 75:1117. https://doi.org/10.1007/s12665-016-5887-8
9. Abiye TA, Sulieman H, Ayalew M (2009) Use of treated wastewater for managed aquifer recharge in highly populated urban centers: a case study in Addis Ababa, Ethiopia. Environ Geol 58:55–59. https://doi.org/10.1007/s00254-008-1490-y
10. Al-Maktoumi A, El-Rawy M, Zekri S, Abdalla O (2015) Managed aquifer recharge using treated wastewater to mitigate seawater intrusion along the Jamma coastal aquifer, Oman. In: International Congress of the International Water Association. International Congress of the International Water Association, Amman, Jordan
11. Renganayaki SP, Elango L (2013) A review on managed aquifer recharge by check dams: a case study near Chennai, India. IJRET 2(4):416–423. https://doi.org/10.15623/ijret.2013.0204002
12. Sherif M, Sefelnasr A, Rashed MA, Alshamsi D, Zaidi FK, Alghafli K, Al-Turbak A, Loni HAOA, Ahamed MB, Ebraheem AA (2023) A review of managed aquifer recharge potential in the Middle East and North Africa Region with examples from the Kingdom of Saudi Arabia and the United Arab Emirates. Water 15(4):742. https://doi.org/10.3390/w15040742
13. Ulibarri N, Garcia NE, Nelson RL, Cravens AE, McCarty RJ (2021) Assessing the feasibility of managed aquifer recharge in California. Water Resour Res 57(3):e2020WR029292. https://doi.org/10.1029/2020WR029292
14. El-Rawy M, Zlotnik VA, Al-Raggad M, Al-Maktoumi A, Kacimov A, Abdalla O (2016) Conjunctive use of groundwater and surface water resources with aquifer recharge by treated wastewater: evaluation of management scenarios in the Zarqa River Basin, Jordan. Environ Earth Sci 75:1146. https://doi.org/10.1007/s12665-016-5946-1
15. Al-Maktoumi A, Zekri S, El-Rawy M, Abdalla O, Al-Abri R, Triki C, Bazargan-Lari MR (2020) Aquifer storage and recovery, and managed aquifer recharge of reclaimed water for management of coastal aquifers. Desalin Water Treat 1(176):67–77. https://doi.org/10.5004/dwt.2020.25499
16. Nsubuga FNW, Namutebi EN, Nsubuga-Ssenfuma M (2014) Water resources of Uganda: an assessment and review. J Water Resource Prot 6:1297–1315. https://doi.org/10.4236/jwarp.2014.614120
17. Kourgialas NN (2021) A critical review of water resources in Greece: the key role of agricultural adaptation to climate-water effects. Sci Total Environ 775:145857. https://doi.org/10.1016/j.scitotenv.2021.145857
18. Ngene BU, Nwafor CO, Bamigboye GO, Ogbiye AS, Ogundare JO, Akpan VE (2021) Assessment of water resources development and exploitation in Nigeria: a review of integrated water resources management approach. Heliyon 7(1). https://doi.org/10.1016/j.heliyon.2021.e05955
19. Miller JE (2003) Review of water resources and desalination technologies. Sandia National Laboratories, California. https://www.osti.gov/servlets/purl/809106
20. Ahmed CB, Rouina BB (2019) Tunisia: water resources' management and sustainability of agriculture. In: Encyclopedia of environmental health, 2nd edn, pp 170–184
21. Allen J, Calvin H (1987) Oman: the modernization of the sultanate. Routledge
22. Charabi Y (2013) Projection of future changes in rainfall and temperature patterns in Oman. J Earth Sci Clim Change 4:154. https://doi.org/10.4172/2157-7617.1000154
23. Khan S, Yang S, Bingchen X, Bradley D, Ahmed N, Shengchun P (2019) Subsurface Highly Saline Water Variability in the Gulf of Oman. Oceans 1–4. https://doi.org/10.1109/OCEANSE.2019.8867120
24. Al-Mashaikhi K, Oswald S, Attinger S, Büchel G, Knöller K, Strauch G (2011) Evaluation of groundwater dynamics and quality in the Najd aquifers located in the Sultanate of Oman. Environ Earth Sci 66:1195–1211. https://doi.org/10.1007/s12665-011-1331-2

25. Diam (2019) Annual Report. Diam. 94 pp. Sultanate of Oman. https://www.diam.om/getattachment/8e37aa3d-b157-46b3-8b2b-75023afff72a/Annual-Report-2019
26. Purnama A, Al-Barwani HH, Al-Rawahi MS, Al-Harthi TM (2004) Dispersion models for brine discharges from desalination plants of Oman. Kuwait J Sci Eng 31(1):1–13
27. Oman Observer (2017) Haya uses advanced technology for wastewater treatment. https://www.omanobserver.om/article/68270/Local%20front/haya-uses-advanced-technology-for-waste-water-treatment
28. Al-Nasri A-M, Gunawardhana LN, Al-Rawas GA, Baawain MS, Sana A (2021) Multi-layer groundwater flow simulation in Al-Khoud lower catchment in Oman. Appl Water Eng Res 10(3):250–260. https://doi.org/10.1080/23249676.2021.1982027
29. Miura M, Arai S, Ahmed AH, Mizukami T, Okuno M, Yamamoto S (2012) Podiform chromitite classification revisited: a comparison of discordant and concordant chromitite pods from Wadi Hilti, northern Oman ophiolite. J Asian Earth Sci 59:52–61. https://doi.org/10.1016/j.jseaes.2012.05.008
30. Zekri S, Ahmed M, Chaieb R, Ghaffour N (2014) Managed aquifer recharge using quaternary-treated wastewater: an economic perspective. Int J Water Resour Dev 30(2):246–261. https://doi.org/10.1080/07900627.2013.837370
31. El-Rawy M, Al-Maktoumi A, Zekri S, Abdalla O, Al-Abri R (2019) Hydrological and economic feasibility of mitigating a stressed coastal aquifer using managed aquifer recharge: a case study of Jamma aquifer, Oman. J Arid Land 11:148–159. https://doi.org/10.1007/s40333-019-0093-7
32. Al-Maktoumi A, El-Rawy M, Zekri S (2016) Management options for a multipurpose coastal aquifer in Oman. Arab J Geosci 9:636. https://doi.org/10.1007/s12517-016-2661-x
33. Ebrahim GY, Jonoski A, Al-Maktoumi A, Ahmed M, Mynett A A (2016) Simulation-optimization approach for evaluating the feasibility of managed aquifer recharge in the Samail Lower Catchment, Oman. J Water Resour Plann Manag 142(2). https://doi.org/10.1061/(ASCE)WR.1943-5452.0000588
34. Al-Rashed MF, Sherif MM (2000) Water resources in the GCC countries: an overview. Water Resour Manag 14:59–75. https://doi.org/10.1023/A:1008127027743

Assessment of Riverbank Filtration (RBF) for Drinking Water Supply in Upper Egypt

Mohamed Galal Eltarabily, Mustafa El-Rawy, and Peter H. Riad

Abstract Riverbank filtration has long been used to provide drinking water to riparian societies around the world. Although this success and its appeal as a low-cost technology, it has not been widely applied as a viable means for water supply in developing countries. Riverbank filtration can provide the added benefit of natural filtration in addition to the enhancement of treatment efficiency even if treatment plants are used in polluted reaches of the rivers. Moreover, for ephemeral rivers or Wadis, prospective applications of riverbank filtration, as well as the storage and recovery of aquifers, might be envisaged. Several ideas related to riverbank filtration, as well as their relevance to desert environments, were discussed in this work. Discussion of the hydrogeochemical processes involved in riverbank filtration and artificial recharging of contaminated surface water, including quantification, identification, and zonation. The riverbank filtration system for water supply along the Nile River (Nile Valley of Egypt) is also presented and evaluated. The study recommends policymakers endeavor to expand the use of Riverbank Filtration (RBF) technology as a strategic move to improve the quality of Nile water and meet the growing demand for drinking water. Furthermore, for the efficient implementation of RBF technology in Egypt, a complete hydrogeological site study is necessary, considering geological, hydrological, and environmental factors to establish optimal site selection.

M. G. Eltarabily (✉)
Civil Engineering Department, Faculty of Engineering, Port Said University, Port Said 42523, Egypt
e-mail: eng_m_trabily@eng.psu.edu.eg

M. El-Rawy
Civil Engineering Department, Faculty of Engineering, Minia University, 61111 Minia, Egypt

Civil Engineering Department, College of Engineering, Shaqra University, DawadmiAr Riyadh 11911, Saudi Arabia

M. El-Rawy
e-mail: mustafa.elrawy@mu.edu.eg

P. H. Riad
Irrigation and Hydraulics Department, Faculty of Engineering, Ain Shams University, Cairo, Egypt
e-mail: peter_riad@eng.asu.edu.eg

M. El-Rawy and A. Negm (eds.), *Managed Aquifer Recharge in MENA Countries*, Earth and Environmental Sciences Library,
https://doi.org/10.1007/978-3-031-58764-1_7

Keywords Riverbank filtration · Water pollution · Drinking water · Groundwater · Water supply · Nile Valley · Egypt

Abbreviations

AR	Artificial Recharge
ASR	Aquifer Storage Recovery
ASTR	Aquifer Storage Transfer Recovery
ETC	Electron Trapping Capacity
MAR	Managed Aquifer Recharge
OMP	Organic Micropollutants
ORP	Oxidation-Reduction Potential
RBF	Riverbank Filtration
SAT	Soil Aquifer Treatment
STP	Sewage Treatment Plant

1 Introduction

In Egypt, the scarcity of water resources presents a formidable challenge, particularly considering the nation's dependence on the Nile River as a primary water source [1]. With a burgeoning population and increasing industrial demands, ensuring the availability of high-quality drinking water has become imperative [2]. Riverbank Filtration (RBF) technology emerges as a strategic initiative to augment the quality of Nile water and address the escalating need for potable water in Egypt. Identified as a natural filtration method, RBF utilizes a river's bed and bank areas as "treatment" zones, where wells, when strategically placed and activated, facilitate the removal of surface water pollutants. With significant potential in densely populated regions globally, this established technology is particularly suited for use in arid countries along permanent or ephemeral rivers. Demonstrating efficiency and cost-effectiveness when managed appropriately, RBF leverages the inherent dynamics of riverbanks, offering an environmentally friendly solution to meet the escalating demand for clean water [3]. RBF units exhibit lower capital and operational expenses than traditional water treatment plants [4]. The diverse applications of Riverbank Filtration span municipal water supply systems to fulfill industrial and agricultural water needs, making it an attractive choice for regions grappling with water scarcity, contamination challenges, or seeking sustainable alternatives to conventional water treatment methods [5].

The feasibility of RBF approach is widely investigated over the world, such in Hungary [6], China [7], India [8], USA [9, 10], Brazil [11], Korea [12], Germany [13] and Egypt [4, 5, 14–16], with diverse hydrological and environmental conditions.

In Europe, RBF is used for providing communities with drinking water while in the USA; it is used for industrial and municipal activities [1].

According to Ghodeif et al. [16], RBF provide a total of 0.1% to Egypt's national drinking water source. Hegazy et al. [15] used RBF to enhance the water quality of the Nile River through the removal of heavy metals. The experimental findings revealed average percentages of heavy metal removal, namely 74.04% for iron, 74.44% for cobalt, 70.72% for lead, 75.1% for zinc, and 70.8% for copper. Abdelrady et al. [5] explored the viability of applying the BF technique in Aswan, Egypt, considering diverse environmental conditions. The study on water quality revealed that the implementation of BF in Aswan City proves to be a successful approach in eliminating chemical contaminants and ensuring the provision of high-quality drinking water.

Riverbank filtration is an alternative to conventional surface water treatment since it utilizes the riverbed and underlying aquifer as natural filters for the removal of contaminants. The process is simple where the drawdown cone causes the water infiltration and movement towards the well when pumping starts (Fig. 1). The most of dissolved solids and suspended pollutants are removed from infiltrating water as it flows through the soil which are typically hard to be removed in treatment plants. To ensure the long-term operation of RBF systems, it's essential to conduct researches to determine the connectivity between surface and groundwater aquifers, the dynamic variation of redox zone(s), the influence of flash floods, and the climate change effects.

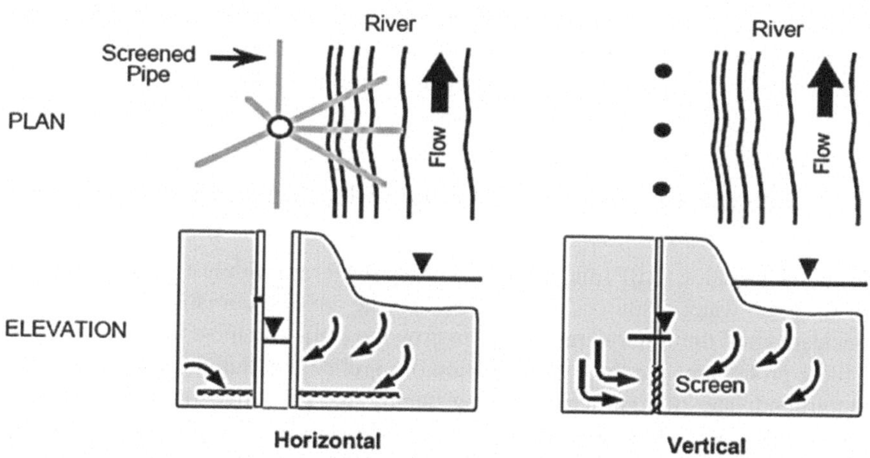

Fig. 1 Conceptual description of an RBF system with horizontal collector and vertical wells (after Ray et al. [10])

2 Concepts and Applicability of Riverbank Filtration

When designing RBF systems, it is fundamental to take into consideration the distance of the well from the river, the pumping rate, the orientation of the well (vertical or horizontal), well depth, and riverbed conditions. These factors affect the water flow and velocity and therefore the travel time from the river to the well. The travel time of water is longer for wells that are far from the river than near wells. Consequently, the travel time of water in an RBF system employing vertical wells is longer than that employing horizontal collector wells. A longer travel time enables the removal of pathogens and the majority of chemicals [10]. However, the dissolution of iron, manganese, and other redox-sensitive chemicals may occur when a long travel distance exists.

Additionally, the design of RBF systems substantially depends on the site's geological and hydrological conditions [5]. For example, the river's flow is flashy and little during the dry seasons in desert countries thus, the wells' yield is low. In these cases, horizontal collector wells should be built inside the river so that the well can always be surrounded by water, even in the dry seasons. Filtration would be enhanced through dam building around the wells site during dry seasons.

Surface water quality depends on the temperature wherein desert environments (with high temperatures), organic carbons are more frequent than in normal temperate regions. When surface water contains a considerable amount of organic carbon (typically expected in developing countries), a fast depletion of dissolved oxygen occurs, therefore causing an anoxic environment. Subsequently, in the desert regions, the organic pollution load is high during low-flow conditions because the dilution is low, considered the most critical period during the water infiltration process [17].

3 Riverbank Filtration and Artificial Recharge Types

Riverbank filtration (RBF) and artificial recharge (AR) are widely used for mitigating the stress on water resources especially when the adverse climate change effect is considered and the water demands are in growing [18]. The most important aspects of these procedures are the transformation of unreliable polluted surface water into safe groundwater of significantly higher quality and the provision of subterranean storage that protects the water from evaporation losses [19]. The ponderous clogging phenomenon [20], water losses due to mixing with brackish groundwater, and natural interactions with soil minerals are all disadvantages of the RBH system. According to the classification shown in Fig. 2, there are several different types of MAR systems [21–28]. These MAR techniques are among the most frequently used for different water supplies, schematized in Fig. 3.

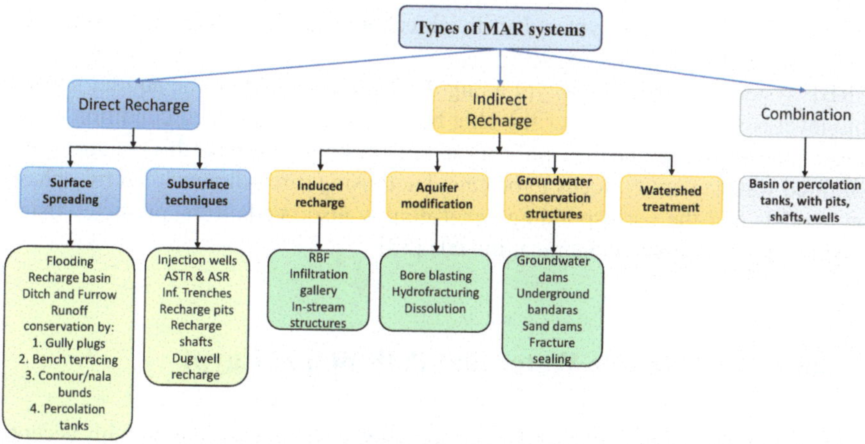

Fig. 2 Classification of MAR systems Modified after Chadha [29]

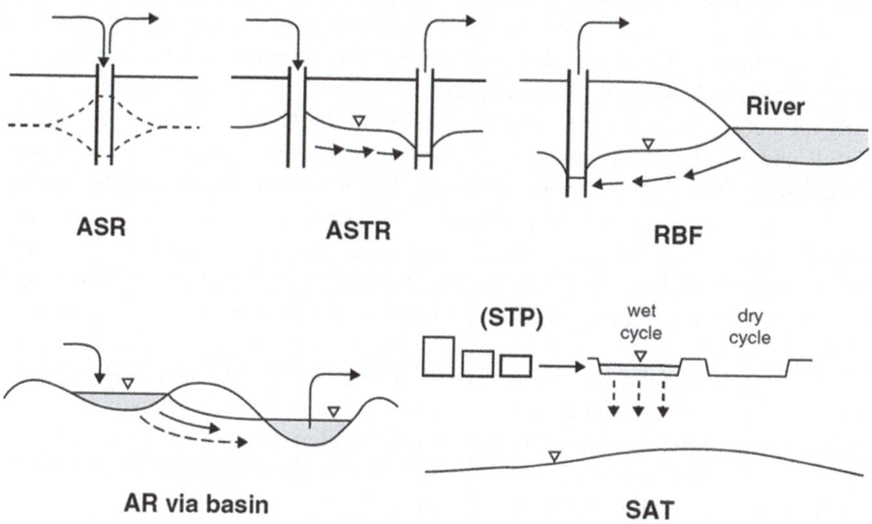

Fig. 3 Schemes of the most common MAR types. Modified after Dillon [30]. (ASR) Aquifer storage recovery; (ASTR) Aquifer storage transfer recovery; (RBF) Riverbank filtration; (AR) Artificial recharge; (SAT) Soil aquifer treatment; (STP) Sewage treatment plant

4 Applications of MAR in Egypt

Managed aquifer recharge (MAR) applications are well-studied by many researchers in Egypt either using physical or numerical models [4, 16, 31–37]. However, these applications are still limited on the real ground, as there are few sites for RBF along the Nile Valley in Upper Egypt and artificial recharge ponds for flash floods retention in Sinai and in the Red Sea coastline.

RBF system is applied in Egypt along the Nile River. When the aquifer's water table level is significantly lower the Nile bed, a partially saturated condition exists between surface water and the aquifer. If an RBF system is built for this condition, the infiltrated water quantity needs to be calculated using three-dimensional two-phase flow models and the corresponding water quality also needs to be predicted using mass transport and biogeochemical models [38].

5 Sources of Potable Water and Pollution in Egypt

In developing countries such as Egypt, meeting water requirements with acceptable quality is becoming exceedingly challenging. Despite all pollution control measures, Egypt's Nile water is becoming increasingly polluted. Industrial and domestic wastewater discharge, reverse drainage of agricultural water, and flash floods are all regarded as major sources of pollution in the Nile River [39]. Current and prior findings of water quality obtained by RBF demonstrate the technology's ability to treat Nile water to a high quality.

Egypt's conventional water supply is limited to the Nile, groundwater, seasonal rainfall, and flash floods. Maintaining the nation's development requires protection of this limited amount of freshwater [40, 41]. Water quality issues associated with these sources vary according to flow, pattern of usage, population density, degree of industrialization, sanitation system availability, and social and economic situations. Furthermore, in Egypt, there is no national monitoring program concerning identifying and determining organic micropollutants in drinking water resources [42].

Pollution load in the Nile system (Nile River, canals) has increased in recent decades consequently, the quality of Nile water has worsened dramatically [43]. The Nile's most polluted section from Cairo to the Mediterranean Sea is within the two branches of the Nile, Damietta, and Rosetta [44, 45]. Figure 4 shows the Nile River and investigated sites for riverbank filtration in Egypt.

Discharging the drainage water into the Nile River represents the main source of pollution. The volume of drainage water is about 12 billion m^3/year, 25% of this volume is in upper Egypt (Nile Valley) and 75% is existing in the Nile delta. There are 76 drains discharging water into the Nile system [46]. The Nile receives discharges from 67 agricultural drains from Aswan to Delta Barrage. Industrial wastewater is considered the second main source of Nile water pollution. The third major source of contamination of the Nile is the effluent of municipal wastewater. About 239 wastewater treatment plants have an annual effluent of 4.5 billion m^3, of which 1,300 million m^3/year are discharged into the Nile River system [14].

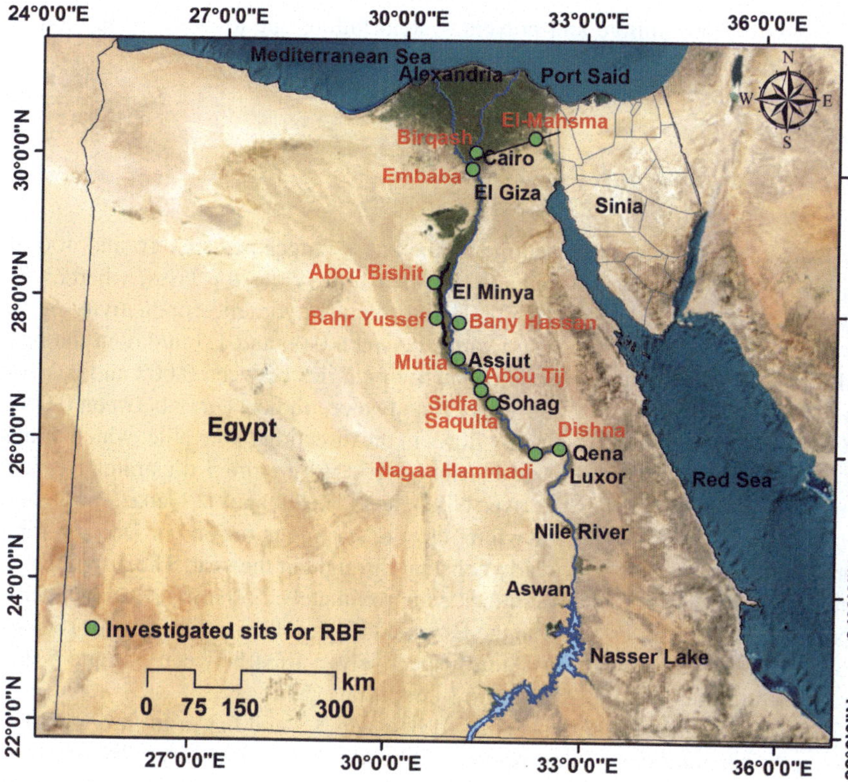

Fig. 4 Map of Egypt shows the Nile River and investigated sites for riverbank filtration. *Source* Created by the authors

6 RBF for Nile Valley Water Supply

In Egypt, RBF attempts to meet the strategic objective of providing high-quality drinking water at an affordable price. Although the use of RBF is still less than 0.1% in Egypt, it could be expanded [16]. Many RBF initiatives have been successfully implemented along the Nile in Upper Egypt, receiving a good share of bank filtration and producing water that meets Egyptian drinking water standards for organic, inorganic, and microbiological characteristics [14, 47–51]. The potential for riverbank development in three Nile Valley locations is assessed using both water level and quality monitoring.

Some of these sites were effective, but others failed to deliver a significant volume of bank filtrate or operate at very high abstraction rates for an extended period due to canal bed clogging. The primary causes of failure were insufficient hydrogeological investigations, improper well design, and/or the absence of operational rules and procedures. Site hydrogeology and river hydrology should be considered to increase the amount of bank filtrate and prevent system failure due to clogging in modern RBF systems [16].

6.1 Nile Water and RBF

The Nile Valley's Quaternary sediments thickness ranges between 16 and 400 m, with an average of 125 m. It has a 20 m-thick impervious top layer, whereas the Nile Delta's is up to 50 m thick [52]. The horizontal hydraulic conductivity of the top cap layer (Holocene sediments) varies between 0.06 and 1.5 m/day in the Nile Valley, while the vertical hydraulic conductivity varies between 0.002 and 0.04 m/day [53, 54]. The Nile Valley aquifer's groundwater gradients vary between 0.0008 and 0.007, and groundwater typically flows in the direction of the Nile, which drains the aquifer and acts as a discharge barrier [5, 55, 56]. Figure 5 illustrates the flow from High Aswan Dam and the river Nile's level (i.e., stage) at Qena. The lowest water level of Nile occurs during winter closure starting at the end of Dec and for 1 month long. The Nile stage is nearly constant throughout the year. The difference in water level between winter and summer is approximately 2.50 m. The average water depth along three profiles between the Esna and Nagaa Hammadi barrages (166 and 346 km downstream of Aswan Dam, respectively) varied between 5.0 and 8.0 m during the minimum and maximum flow conditions [57].

The Nile River's typical depths varied between 3.3 and 6.3 m in three sections (Luxor, Aswan, and Qena), and between 2.3 and 5.0 m in El-Minya. The depth of the river has been highlighted as a significant characteristic for RBF implementation. RBF is not practicable if the depth of the river is less than the clay cap thickness [58].

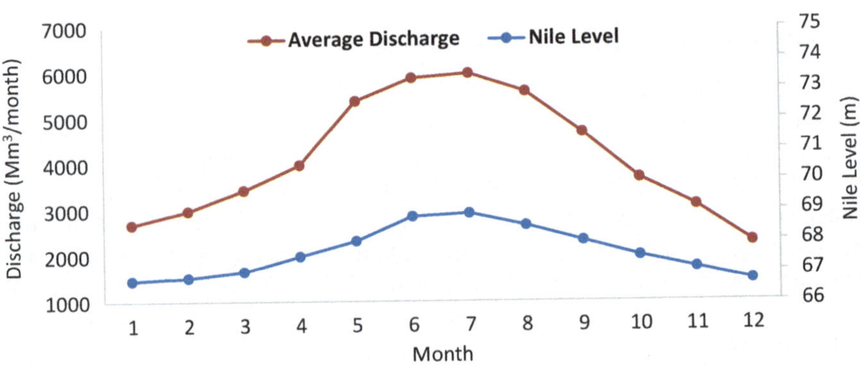

Fig. 5 Average monthly discharge of the Nile River and the water level, 2008. Modified after Shamrukh and Abdel-Wahab [14]

6.2 Nile-Aquifer Interaction

To recover bank filtrate, a strong hydraulic relationship between the river and the aquifer is required. The Nile riverbed must either cut into the aquifer or be deeper than the top layer's bottom edge. Hydrogeological profiles in Upper Egypt (Fig. 6) at El-Minya and Abou Tij show full penetration of the Nile through the clay cap layer while Dishna shows partial penetration. Values in the Nile valley have a moderate hydraulic gradient ranging from 0.008 to 0.0001. The groundwater flow at the main aquifer is towards the Nile River. The main aquifer is mainly composed of deposits such as gravel and sand that have relatively high hydraulic conductivity values that range from 75 to 100 m/d [59].

Nile River and adjacent groundwater levels were continuously monitored from Oct. 2012 to Nov. 2013 (Fig. 7) to investigate the effect of seasonal changes in the River Nile's water level on the groundwater flow regime. Summer is a season when shallow aquifers have the most water. Water level variations over time are comparable for both the Nile and the aquifer, implying a direct hydraulic relationship. The Nile valley has more suitable hydrogeological conditions for RBF applications than the Nile delta [37]. The Nile River cuts through the clay cap completely or partially, and

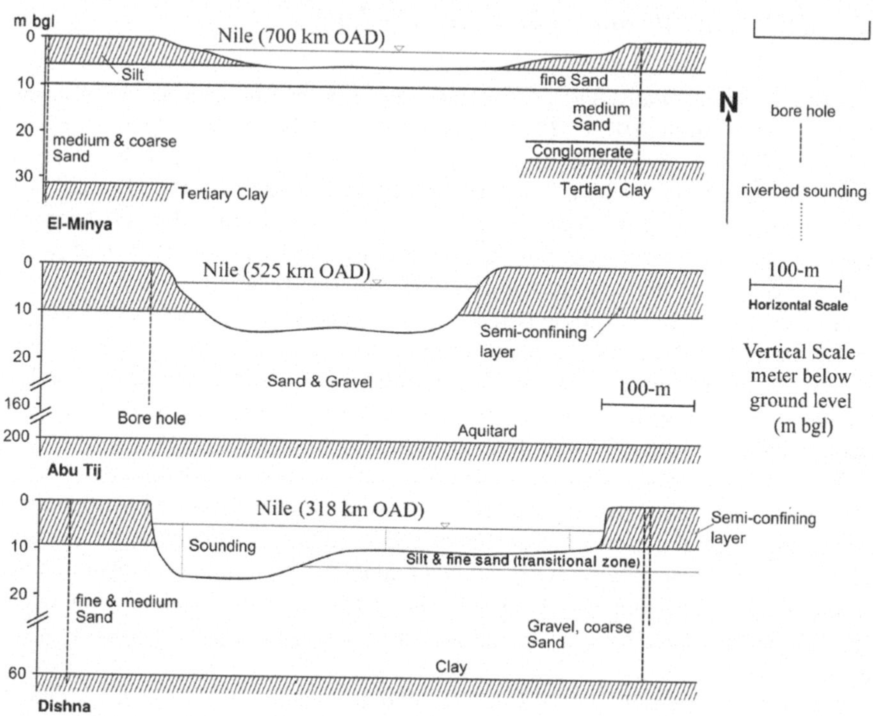

Fig. 6 Nile Valley's hydrogeologic cross-sections at El-Minya, Abu Tij, and Dishna (Ghodeif et al. [16])

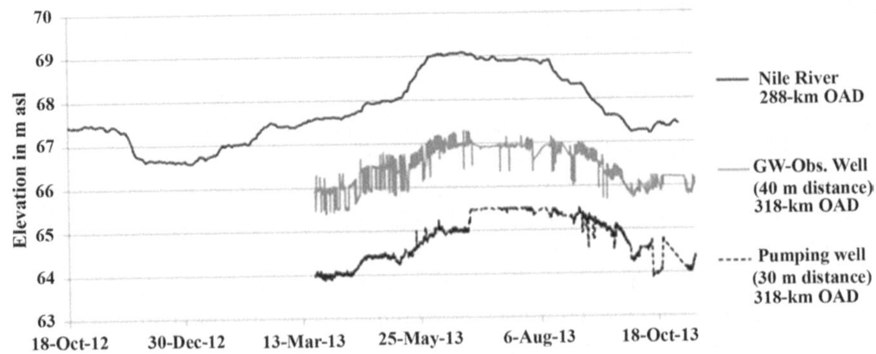

Fig. 7 River Nile is compared to the changes in groundwater levels in Dishna, Qena. Modified from Bartak et al. [60]

so is hydraulically connected to the Quaternary aquifer, whose riverbed materials are sand [2].

6.3 RBF Benefits and Scheme in Nile Valley

The average water quality indicators values for both the Nile and its major canals in Upper Egypt from 2012 to 2014 are presented by Ghodeif et al. [16], as shown in Fig. 8. The quality of the River Nile's water is detailed at three locations in Upper Egypt namely, Dishna, Abou Tij, and El-Minya, respectively 318, 525, and 700 km downstream of the Aswan Old Dam (AOD). The average dissolved oxygen concentration in Upper Egypt ranges from 8.2 mg/l in Dishna to 10.2 mg/l in El-Minya. Because of the salinity increase downstream of the Nile River, the mean electric conductivity (EC) of the Nile River varies between (315 and 394 μS/cm) as well as the mean chloride (Cl^-) concentrations vary between 17 and 23 mg/l. Water in the main canals has the same chemical composition as Nile water. Turbidity is modest but above drinking water standards, reaching 8.8 NTU at EL-Minya. Turbidity can increase by a 100-fold during flash floods [16].

The concentrations of all main ions in Nile water are normally within the permissible limits for drinking water. The Nile water is alkaline and mostly bicarbonate in composition. Alkalinity is between 160 and 169 mg/l, with calcium as the primary hardness component. The primary source of contamination is an excessive number of microbiological pollutants. At Dishna, the mean total coliform (TC) numbers of 2819 MPN/100 ml and the total algae count of 6118 org/ml surpassed drinking water standards [61].

Nile Valley's ambient groundwater, close to the existing Nile River track, has an EC value of between 344 and 1510 μS/cm, with an average of 875 μS/cm (Table 2). The average values of (Iron, Fe) and (Manganese, Mn) are 0.43 mg/l and 0.49 mg/l, respectively. (Fe) concentrations vary from <limit of quantification (LOQ) (0.06:1.05

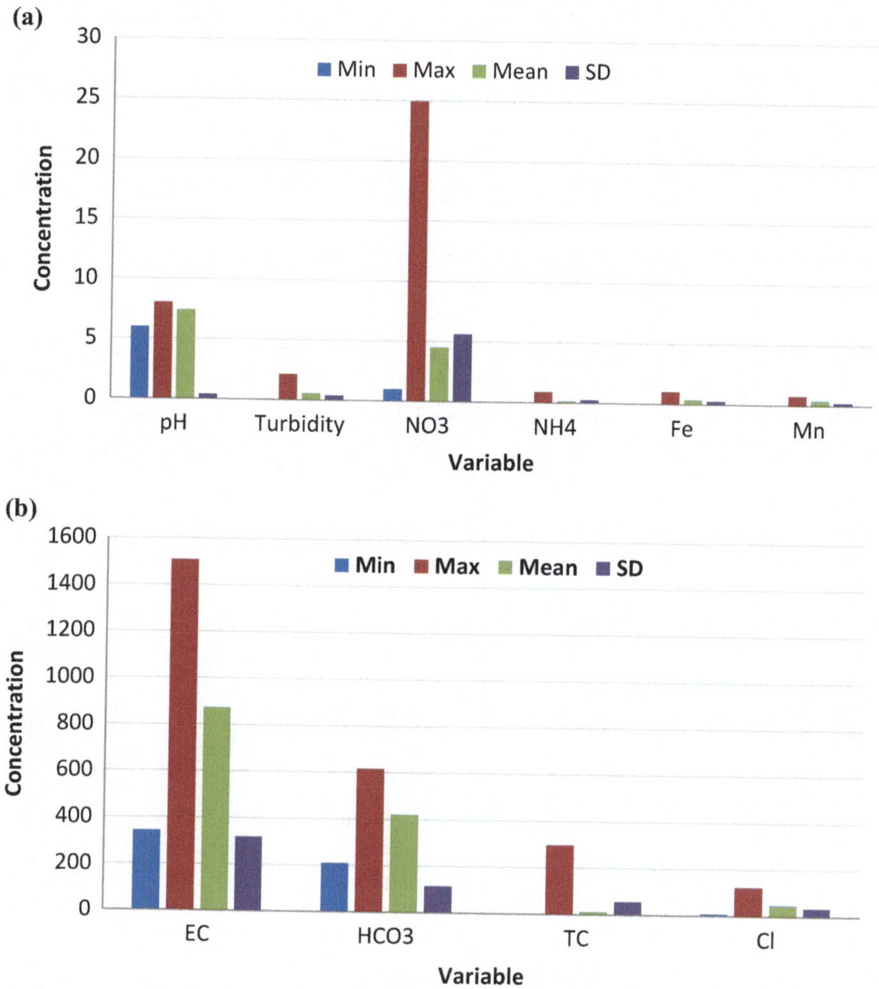

Fig. 8 Statistical analysis of the water quality at Qena drinking water wells (2013/2014): **a** pH, Turbidity, NO_3, NH_4, Fe, and Mn; **b** EC, HCO_3, TC, and Cl

mg/l), whereas (Mn) concentrations range from <LOQ (0.04:0.84 mg/l). Ammonium $\left(NH_4^-\right)$ and nitrate $\left(NO_3^-\right)$ had mean values of 0.20 mg/l and 4.53 mg/l, respectively. Mean Values for ammonium $\left(NH_4^-\right)$ and nitrate $\left(NO_3^-\right)$ are 0.20 mg/l and 4.53 mg/l. The ammonium and nitrate have maximum contents of 0.94 mg/l and 25 mg/l, respectively. Total coliform (TC) values surpassed drinking water regulations in several wells. In the Nile valley aquifer, it has a mean value of 14 MPN/100 ml.

6.4 Evaluation of Bank Filtration Sites Within 2012–2014

Bank filtration sites were assessed in terms of bank filtration share and the quality of produced water. The bank filtration share (BF%) was determined by averaging the values of conservative species (chloride and EC) (Fig. 9). The following equation was used to determine the percentage of bank filtrate (BF %) in the pumped water:

$$BF\% = \frac{C_{sample} - C_{Groundwater}}{C_{Surfacewater} - C_{Groundwater}} \times 100\% \quad (1)$$

where BF % is bank filtrate share in %; *C* concentration (or EC) of conservative species used; *C Sample* concentration (or EC) in the pumped water; *C Groundwater* concentration (or EC) in groundwater; C surface water concentration (or EC) in surface water.

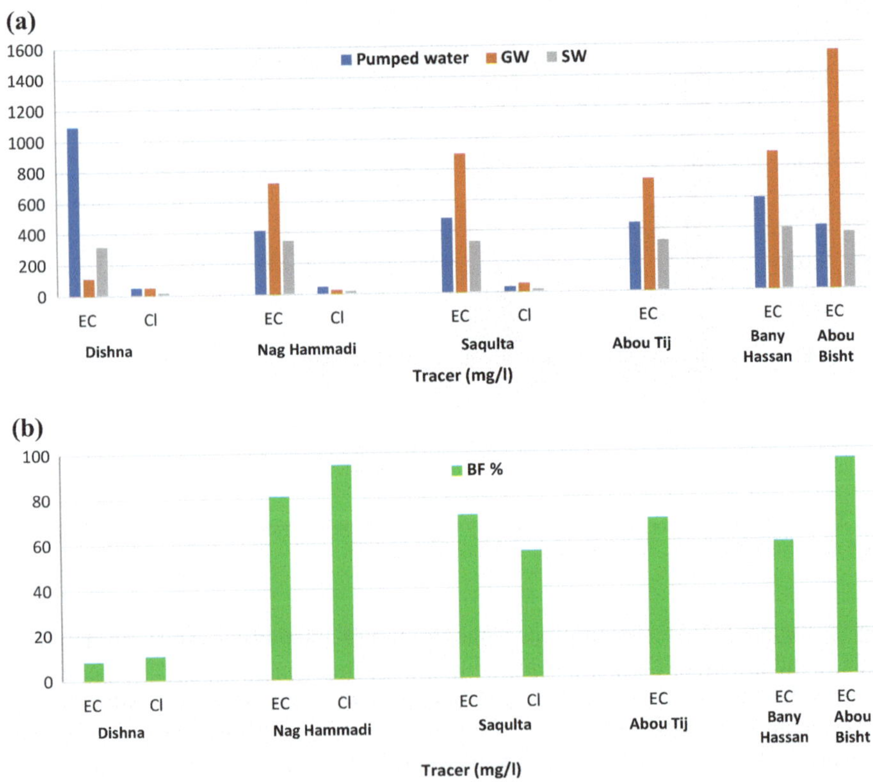

Fig. 9 **a** EC and Cl concentration in the pumped water, Groundwater (GW), and Surface water (SW); **b** bank filtration share (BF%) at the investigated sites. Modified after Ghodeif et al. [16]

The RBF sites along the Nile in Upper Egypt that have a BF greater than 50% indicate the feasibility of RBF. Two sites (Dishna and Naga Hammadi) are located along the eastern bank of the River Nile in Qena governorate; one was able to deliver approximately 80% bank filtrate while the other has failed to deliver >10% so far. These sites produce drinking water that meets the permissible limit of Egyptian drinking water standards [16].

Canal bed clogging is generally more common than riverbed clogging on the Nile River, as flow velocities are lower. Canals are frequently prone to silt and clay sedimentation and are dredged every three months. This mud layer has the potential to drastically reduce penetration rates and increase iron and manganese release. Alluvial sediments in the Nile River valley have a high hydraulic conductivity, allowing for the placement of wells at varying distances from the riverbank. However, the natural groundwater flow gradient toward the river appears to be relatively high for many potential RBF sites. As a result, abstraction wells should be located close to the riverbank to ensure adequate drawdown and a high portion of bank filtrate [16].

Despite Egyptian regulations requiring the drilling of wells for potable supply to a depth of 60 m (Egyptian Ministry of Health and Population decree No. 1 1994, High Water Committee decisions year 1997), RBF wells should be designed to extract water from the aquifer's upper layer in order to increase bank filtrate. It is critical to maintain continuous operation of the pumps in the wells. Discontinued pump operation considerably reduces bank filtrate infiltration into the aquifer and, consequently, the percentage of bank filtrate in the final abstracted water, resulting in water quality problems (iron, manganese, and ammonium).

6.5 RBF Investigations in Three Sites in Upper Egypt from 2018 to 2021

In 2018 many new RBF wells were constructed at three production sites in Upper Egypt. These sites are Alsaayda in Luxor, Almonsha in Sohag and Abou Tij in Assiut (Fig. 10). Six RBF wells were constructed in Alsaayda site in the north of Luxor city. These wells' main function is covering Alzeneya district needs from drinking water, as the population in this area were suffering from water shortage. For the same reason, three RBF wells were constructed in each site in Almonsha in Sohag and Abou Tij in Assiut. Water abstracted from all wells is disinfected by chlorine in water storage tanks before being pumped to the network [62].

The soil log formation in each site was investigated as shown in Fig. 11 [62]. From these formations it was noticed that the first and third sites have shallow clayey caps (within 2 m thickness), while the clayey cap in Almonsha, Sohag has deeper thickness reaches to 6 m. Table 1 gives an overview about all RBF wells at the three sites.

Water sampling from RBF sites and the Nile River were taken and analyzed according to the Egyptian Guidelines for Drinking Water and Standard Methods for

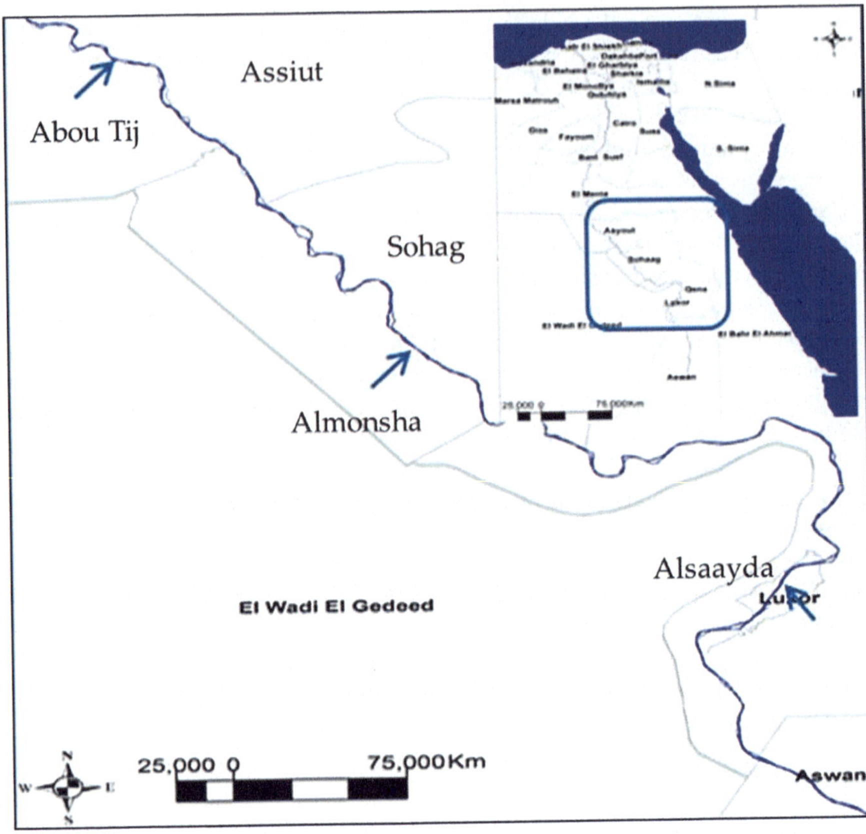

Fig. 10 Sites locations for RBF investigations within 2018–2021 (after [62])

Water and Wastewater [63, 64]. A series of continuous monitoring were taken from all sites.

6.6 *Evaluation of Bank Filtration Sites from 2018 to 2021*

Table 2 gives all average values of major water quality parameters for Nile water and pumped water. EC in pumped water from Sohag site is higher than EC in The Nile River; this is due to the high salinity of the ambient groundwater, as this area has agricultural activities that use fertilizers intensively. This, in turn, raised the concentration of chloride and sulfate, too.

Low chloride concentrations at Luxor and Assiut indicate a high efficiency of bank filtration from The Nile River. This can be interpreted due to the riverbed penetration below the shallow clayey caps in these sites. pH is higher in surface water than in the pumped water at all sites because of degradation processes in the aquifer and the

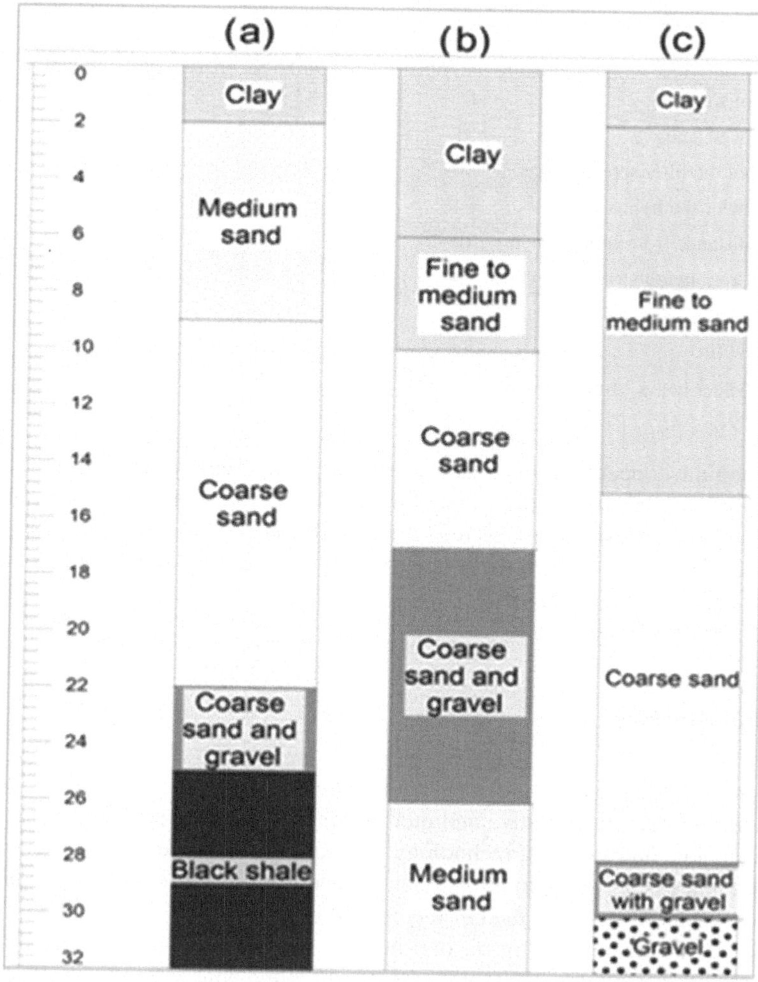

Fig. 11 Soil log formation for sites. **a** Alsaayda in Luxor, **b** Almonsha in Sohag and **c** Abou Tij in Assiut (after [62])

formation of CO_2. CO_2 is further reacting with the minerals in the aquifer, resulting in increased alkalinity (HCO_3) [62].

Fe concentrations at all three sites are below the threshold for drinking water of 0.3 mg/l. In Sohag and Luxor, Mn concentrations are above the drinking water threshold of 0.4 mg/L, whereas in Assiut Mn concentration is 0.29 mg/l only. From previous investigations of the initial phase of RBF, it is known that the washout effect in the aquifer takes longer time for Mn than for Fe [4]. It is important to point out that Fe and Mn concentrations were decreased with continuous pumping. Post-treatment, like using green sand filters, can be used for reducing Mn and Fe concentrations, if they are persist.

Table 1 RBF wells characteristics in each site

Site	Alsaayda, Luxor	Almonsha, Sohag	Abu Tieg, Assiut
Number of wells	6	3	3
Depth of well (mbgs)*	31	36	36
Location of the filter screen (mbgs)	10–25	18–34	18–34
Well diameter (inch)	14	12	12
Average distance from riverbank (m)	10–13	15	20
Distance from neighboring well (m)	19–25	15	12
Pumping rate (m^3/h)	108	108	108
Static level (m)	4.6	7.2	5.7

* mbgs—Meter below ground surface

Ammonium concentrations are also increasing compared to river water but remain on average below the threshold value of 0.5 mg/l. No coliforms and fecal coliforms were detected in pumped water. Removal of turbidity and pathogens is a major advantage in RBF in all sites. Also, the decrease in colony-forming units of heterotrophic bacteria (HPC) is impressive [62].

7 Conclusions

Riverbank filtration as a natural treatment method could be the optimal solution to overcome Egyptian quantitative and qualitative challenges of drinking water. RBF is a sustainable and low-cost technology that can help. It is not viable everywhere in Egypt; it needs favorable hydrogeological conditions and thorough investigations. Previous research work and data of the present investigation show that RBF can be an effective and economical means of drinking water production from the Nile. RBF can be used as a standalone or as an alternative to direct Nile abstraction followed by a conventional treatment process. Continuous operation of the RBF wells prevents fluctuation in water quality and limits the pumped portion of ambient groundwater which may have higher Fe and Mn concentrations. Fully penetrating sections, islands, and meanders of the Nile River profile are the best sites for RBF applications. In general, RBF can be a substantial source of water with high quality and low cost to cover the water shortage in Upper Egypt.

Table 2 Drinking water quality standards, average water analysis for water from The Nile River, bank filtration (BF) in Luxor, Sohag and Assiut between 2018 and 2021 [62]

Parameter	Unit	Egyptian standards	Nile water	BF Luxor 2018–2021	BF Sohag 2018–2021	BF Assiut 2020–2021
EC*	μS/cm	–	480 (n** = 720)	430 (n = 144)	937 (n = 144)	360 (n = 48)
pH	–	6.5–8.5	8.1 (n = 1080)	7.6 (n = 96)	7.6 (n = 96)	7.5 (n = 48)
Turbidity	NTU	1	3.7 (n = 1080)	0.4 (n = 144)	0.8 (n = 144)	0.3 (n = 48)
TDS	mg/L	1000	307 (n = 720)	275 (n = 144)	600 (n = 144)	230 (n = 48)
Alkalinity	mg/L	–	132 (n = 360)	160 (n = 48)	230 (n = 48)	NA
Total hardness	mg/L	500	115 (n = 360)	NA	NA	NA
Fe*	mg/L	0.3	<0.001 (n = 48)	0.21 (n = 144)	0.28 (n = 144)	0.18 (n = 48)
Mn*	mg/L	0.4	0.01 (n = 48)	0.45 (n = 144)	0.65 (n = 144)	0.29 (n = 48)
NH_4*	mg/L	0.5	0.01 (n = 48)	0.39 (n = 144)	0.41 (n = 144)	0.45 (n = 48)
Cl*	mg/L	250	14 (n = 360)	29 (n = 96)	41 (n = 96)	22 (n = 48)
SO_4	mg/L	250	20 (n = 360)	36 (n = 96)	50 (n = 96)	51 (n = 48)
HPC*	CFU/ml	50	2100 (n = 360)	10 (n = 90)	6 (n = 90)	2 (n = 48)
Total coliform*	CFU/100ml	<1	2400 (n = 360)	0 (n = 72)	0 (n = 72)	0 (n = 48)
Fecal coliform*	CFU/100ml	<1	200 (n = 360)	0 (n = 72)	0 (n = 72)	0 (n = 48)

* Main parameters of bank filtration analysis, NA—Not analyzed
** n = number of samples

8 Recommendations

According to the findings, the Egyptian government should widen the use of Riverbank Filtration (RBF) technology as a strategic step to meet the growing demand for drinking water. For the effective use of Riverbank Filtration (RBF) technology in Egypt, a detailed hydrogeological site assessment is required, taking geological, hydrological, and environmental factors into account to determine optimal site selection. Climate resilience should be incorporated into the design of RBF systems, with the impact of changing climatic circumstances assessed. Implementing thorough water quality monitoring procedures will enable continual evaluation and improvement of RBF efficiency in eliminating common contaminants. Incorporating of RBF into national and regional water management plans is critical in order to match its implementation with broader efforts to meet water demand and maintain resource sustainability.

Economic feasibility studies should be carried out to provide decision-makers with sufficient information on the cost-effectiveness and long-term benefits of RBF in comparison to other water treatment methods. Community engagement measures should be focused on promoting awareness and gathering support, while urban adaptations and the establishment of supportive policies and regulatory frameworks are critical. Capacity-building initiatives and knowledge-sharing platforms must be built to boost local expertise and improve stakeholder engagement, ensuring successful RBF implementation and sustainability in the Egyptian context.

References

1. Negm AM, El-Rawy M (eds) (2022) Sustainability of groundwater in the Nile Valley, Egypt. Springer Nature
2. El-Rawy M, Makhloof AA, Hashem MD, Eltarabily MG (2021) Groundwater management of the Quaternary aquifer of the Nile Valley under different recharge and discharge scenarios: a case study of Assiut governorate, Egypt. Ain Shams Eng J 12(3):2563–2574
3. Shamrukh M, Abdel-Wahab A (2008) Riverbank filtration for sustainable water supply: application to a large-scale facility on the Nile River. Clean Technol Environ Policy 10(4):351–358
4. Wahaab RA, Salah A, Grischek T (2019) Water quality changes during the initial operating phase of riverbank filtration sites in Upper Egypt. Water 11:1258. https://doi.org/10.3390/w11061258
5. Abdelrady A, Sharma S, Sefelnasr A, El-Rawy M, Kennedy M (2021) Analysis of the performance of bank filtration for water supply in arid climates: case study in Egypt. Water 12:1816. https://doi.org/10.3390/w12061816
6. Nagy-Kovács Z, Davidesz J, Czihat-Mártonné K, Till G, Fleit E, Grischek T (2019) Water quality changes during riverbank filtration in Budapest, Hungary. Water 11:302. https://doi.org/10.3390/w11020302
7. Pan W, Huang Q, Huang G (2018) Nitrogen and organics removal during riverbank filtration along a reclaimed water restored river in Beijing, China. Water 10:491. https://doi.org/10.3390/w10040491
8. Boving TB, Patil K, D'Souza F, Barker SF, McGuinness SL, O'Toole J, Sinclair M, Forbes AB, Leder K (2018) Performance of riverbank filtration under hydrogeologic conditions along the upper Krishna River in Southern India. Water 11:12
9. Saxena S, Satsangi A, Soamidas V (2021) River bank filtration system: cost effective water supply alternative. In: Water safety, security and sustainability: threat detection and mitigation, pp 565–579
10. Ray C, Melin G, Linsky RB (2002) Riverbank filtration—Improving source water quality. Kluwer, Dordrecht, Boston, London
11. Romero-Esquivel LG, Grischek T, Pizzolatti BS, Mondardo RI, Sens ML (2017) Bank filtration in a coastal lake in South Brazil: water quality, natural organic matter (NOM) and redox conditions study. Clean Technol Environ Policy 19:2007–2020
12. Noh J-H, So S-H, Park J-W, Kim S-Y, Song K-G, Choi J, Kim G-B, Son H, Kim H, Maeng S-K (2022) An assessment of the effectiveness of riverbank filtration in a sewage plant effluent-impacted river using a full-scale horizontal well. Water 14:1873. https://doi.org/10.3390/w14121873

13. Grischek T, Schoenheinz D, Sandhu C, Eckert P (2009) Sustainability of river bank filtration in Germany. In: Proceedings international conference water, environment, energy and society (WEES-2009), New Delhi, pp 12–16
14. Shamrukh M, Abdel-Wahab A (2011) Water pollution and riverbank filtration for water supply along River Nile, Egypt. Riverbank filtration for water security in desert countries. Springer, Netherlands, pp 5–28
15. Hegazy MH, Essam A, Elnaggar AY, E. Hussein E (2021) Heavy metal removal from the water of the River Nile using riverbank filtration. Water 13:3642. https://doi.org/10.3390/w13243642
16. Ghodeif K, Grischek T, Bartak R, Wahaab R, Herlitzius J (2016) Potential of riverbank filtration (RBF) in Egypt. Environ Earth Sci 75:671
17. Grischek T, Bartak R (2016) Riverbed clogging and sustainability of riverbank filtration. J Water 8:604
18. Gale I (2005) Strategies for managed aquifer recharge (MAR) in semi-arid areas. UNESCO IHP, www.iah.org/recharge, 34 p.
19. Pyne RDG (2005) Aquifer storage recovery: a guide to groundwater recharge through wells, 2nd edn. ASR Press, Gainesville
20. Hubbs SA (2006) Riverbank filtration hydrology; impacts on stream capacity and water quality. In: NATO science series IV, earth, and environmental sciences, vol. 60, Springer, 344 p.
21. Stuyfzand PJ (2002) Quantifying the environmental impact and sustainability of artificial recharge systems. In Dillon PJ (ed) Management of aquifer recharge for sustainability, proceedings of the 4th international symposium on artificial recharge, Adelaide, Australia, Balkema, pp 77–82
22. Stuyfzand PJ (2002) Modelling the accumulation rate and chemical composition of sludges in recharge basins. In: Dillon PJ (ed) Management of aquifer recharge for sustainability, proceedings of the 4th international symposium on artificial recharge, Adelaide, Australia, Balkema, pp 221–224
23. Medema GJ, Stuyfzand PJ (2002) Removal of micro-organisms upon basin recharge, deep well injection and riverbank filtration in the Netherlands. In: Dillon PJ (ed) Management of aquifer recharge for sustainability, proceedings of the 4th international symposium on artificial recharge, Adelaide, Australia, Balkema, pp 125–131
24. Stuyfzand PJ, Kortleve M, Olsthoorn TN, Rolf H (2007) Accumulation of pollutants and aquifer leaching during artificial recharge compared to a natural recharge system, in the Netherlands. In: Fox P (ed) Management of aquifer recharge for sustainability, proceedings of the ISMAR-6, Phoenix AR USA. Acacia Publ. Inc., Phoenix, pp 174–184
25. Houben G, Treskatis C (2007) Water well rehabilitation and reconstruction. McGraw-Hill, New York
26. Van Beek CGEM, Breedveld RJM, Juhasz-Holterman M et al (2009) Cause and prevention of well bore clogging by particles. Hydrogeol J 17:1877–1886
27. El-Rawy M, Zlotnik VA, Al-Raggad M, Al-Maktoumi A, Kacimov A, Abdalla O (2016) Conjunctive use of groundwater and surface water resources with aquifer recharge by treated wastewater: evaluation of management scenarios in the Zarqa River Basin, Jordan. Environ Earth Sci 75:1–21
28. El-Rawy M, Al-Maktoumi A, Zekri S, Abdalla O, Al-Abri R (2019) Hydrological and economic feasibility of mitigating a stressed coastal aquifer using managed aquifer recharge: a case study of Jamma aquifer, Oman. J Arid Land 11:148–159
29. Chadha DK (2002) State of art of artificial recharge applied on village level schemes in India. In: Management of aquifer recharge and subsurface storage; making better use of our largest reservoir. Netherlands National Committee IAH, Publ. No. 4:19–24
30. Dillon P, Toze S (eds) (2005) Water quality improvements during aquifer storage and recovery. American Water Works Association. Research Foundation Report 91056F, 286 p + 2 CDs
31. Abdel-Fattah M, Kantoush SA, and Sumi T (2021) Flash floods mitigation using structural measures in Egypt. In: Towards a sustainable water future, pp 295–310
32. Abdel-Fattah M, Kantoush SA, Mohamed S, Sumi T (2018). Rainfall-Runoff modeling for extreme flash floods in Wadi Samail, Oman. J Jpn Soc Civil Eng Ser B1 (Hydr Eng) 74(5):I_691–I_696, https://doi.org/10.2208/jscejhe.74.5_I_691

33. Alkhadrawy A, Riad P, Hassan N, Abdel Salam M (2017) ArcGIS and numerical model for studying groundwater pollutant transport problem in Sadat City in Egypt. Al-Azhar University Civil Eng Res Mag (CERM) 39(3):341–354
34. Riad P, Billib M, Boochs P, Hassan A, Nour Eldin M, Omar M (2012) Shallow wells and surface spreading basins modeling to augment the groundwater resources in a semi-arid area in Egypt. In: VII international conference on environmental hydrology, Cairo-Egypt
35. Riad P, Billib M, Boochs P, Hassan A, Omar M (2011) Analytical and numerical approaches for the hydraulic design of artificial recharge ponds in a Semi-Arid Area. Franzius-Institut für Wasserbau und Küsteningenieurwesen, Leibniz Universität Hannover, Mitteilungen Heft 98, Seite 163–178, Hannover-Germany
36. Riad P, Billib M, Boochs P, Hassan A, Omar M (2011) Overlay weighted model and fuzzy logic to determine the best locations for artificial recharge of groundwater in a semi-arid area in Egypt. Nile Basin Water Sci Eng J 4(1):24–35
37. Abd-Elaty I, Ghanayem HM, Zeleňáková M, Mésároš P, Saleh OK (2021) Numerical investigation for riverbank filtration sustainability considering climatic changes in arid and semi-arid regions; case study of RBF Site at Embaba, Nile Delta, Egypt. Sustainability 13:1897. https://doi.org/10.3390/su13041897
38. Eckert P (2011) Risk assessment for chemical spills in the River Rhine. In: Ray C, Shamrukh M (eds) Riverbank filtration for water security in desert countries. Springer, Berlin, pp 69–79
39. MWRI (2002) Survey of Nile system pollution sources. APRP-Water Policy Activity, Ministry of Water Resources and Irrigation (MWRI), EPIQ Report No. 64
40. Egyptian Environmental Affairs Agency (EEAA) (2009) Egypt State of Environment Report 2008 (Internal report Ministry of State for Environmental Affairs-Cairo-Egypt)
41. El-Rawy M, Abdalla F, El Alfy M (2020) Water Resources in Egypt. In: Hamimi Z, El-Barkooky A, Martínez Frías J, Fritz H, Abd El-Rahman Y (eds) The geology of Egypt. Regional geology reviews. Springer, Cham. https://doi.org/10.1007/978-3-030-15265-9_18
42. Badawy MI, Emababy MA (2010) Distribution of polycyclic aromatic hydrocarbons in drinking water in Egypt. Desalination 251(1–3):34–40
43. Abdel-Dayem S, Abdel-Gawad S, Fahmy H (2007) Drainage in Egypt: a story of determination, continuity, and success. Irrig Drain J 56:S101–S111
44. Abdo MH (2004) Environmental studies on the River Nile at Damietta Branch region, Egypt. J Egypt Acad Soc Environ Dev 5(2):85–104
45. NAWQAM (2003) Nile Research Institute data. National Water Quality and Availability Management Project (NAWQAM), National Water Research Centre. Egypt
46. World Bank (2005) Arab Republic of Egypt: country environmental analysis 1992–2002. The World Bank, Washington DC
47. Shamrukh M, Abdel-Lah A (2004) Monitoring of agricultural nutrients in water supply wells, Upper Egypt. J Eng Appl Sci Cairo Univ 51(6):1135–1150
48. Abdel-Lah A, Shamrukh M (2006) Riverbank filtration: a promise method for water supply from Nile, Egypt. In: Proceedings of 7th international symposium on water supply technology, Yokohama, Japan, pp 385–395
49. Abdalla FA, Shamrukh M (2009) Riverbank filtration as a new treatment technology in Nile Valley: case study, AbuTieg City, Assiut, Egypt. In: NATO advanced research workshop (Riverbank Filtration for Water Security in Desert Countries), Luxor
50. Abdalla FA, Shamrukh M (2011) Riverbank filtration as a new treatment technology in Nile valley: case study, Abou Tieg City, Assiut, Egypt. In: Ray C, Shamrukh M (eds) Riverbank filtration for water security in desert countries. Springer, Dordrecht, pp 255–266
51. Abdel-Lah AK (2013) Riverbank filtration for water supply in semiarid environment. J Eng Sci Assiut Univ 41(3):840–850
52. Geirnaert W, Laeven MP (1992) Composition and history of groundwater in the western Nile delta. J Hydrol 138(1–2):169–189
53. Warner JW, Gates TK, Mankarious WF (1991) Vertical leakage in Egypt's Nile valley: estimation and implication. J Irrig Drainage Eng 117(4):515–533

54. Ahmed AA, Ali MH (2009) Hydrochemical evolution and variation of groundwater and its environmental impact at Sohag, Egypt. Arab J Geosci 4:339–352
55. Brikowski TH, Faid AM (2006) Pathline-calibrated groundwater flow models for Nile Valley aquifers, Esna, upper Egypt. J Hydrol 324(1–4):195–209
56. Hamdan AM (2013) Hydrogeological and hydrochemical assessment of the quaternary aquifer south Qena City, Upper Egypt. Earth Sci Res 2(2):11–22
57. Ismail SS, Samuel MG (2011) Response of River Nile dredging on water levels. In: 15th Internation Water Technology Conference, IWTC-15, Egyptian Water Technology Association, Cairo, Egypt
58. Caldwell TG (2006) Presentation of data for factors significant to yield from several riverbank filtration systems in the U.S. and Europe. Stephen A. In: Hubbs S (ed) Riverbank filtration hydrology. Springer, pp 299–344
59. El-Rawy M, Moghazy HE, Eltarabily MG (2021) Impacts of decreasing Nile flow on the Nile valley aquifer in El-Minia governorate, Egypt. Alex Eng J 60(2):2179–2192
60. Bartak R, Grischek T, Ghodeif K, Wahaab R (2015) Shortcomings of the RBF Pilot Site in Dishna. Egypt J Hydrol Eng 20(9):05014033
61. EHCW (2007) Egyptian standards for drinking water and domestic uses. Egyptian Higher Committee for Water (in Arabic)
62. Salah A, Saber M, Roushdy A, Atef M, Grischek T, Abel Wahaab R (2021) Riverbank filtration for sustainable drinking water supply in Upper Egypt. In: Proceeding of Cairo Water Week, CWW2021, Ministry of Water Resources & Irrigation, 10 p.
63. ESSPDW (2007) Egyptian standard specifications for Potable drinking water. Decision of the Minister of Health and Population No. 458; Minister of Health and Population: Cairo, Egypt
64. AWWA (2019) American water works association, water environment federation (WEF), American Public Health Association (APHA). Standard Methods for the Examination of Water and Wastewater (SMWW). https://doi.org/10.2105/SMWW.2882.219

A Tale of Two Countries: The Potential of Managed Aquifer Recharge in Pakistan and Egypt

Amar Razzaq, Yewang Zhou, Muhammad Aamir Shahzad, Lianfen Wang, and Moataz Eliw

Abstract Egypt and Pakistan face significant challenges in groundwater management. This chapter investigates groundwater management challenges in Egypt and Pakistan, focusing on the pivotal role of the Indus Basin Irrigation System (IBIS) in Pakistan. Despite its abundant irrigation and groundwater resources, IBIS has faced depletion due to stalled surface storage since 1976, leading to an annual groundwater decline of 1.2 feet. The chapter suggests Managed Aquifer Recharge (MAR) through surface water as a viable solution to rehabilitate IBIS's groundwater. In Egypt, where groundwater use surpasses safe yield, key aquifers are overexploited. The chapter highlights effective supply-side management, especially in the Nile Delta and coastal regions, emphasizing the MAR technique to enhance groundwater. The chapter concludes with urgent recommendations for policymakers in both countries to address these critical groundwater management challenges.

Keywords Groundwater management · Groundwater resources · Pakistan · Egypt · MAR · Water scarcity

---

A. Razzaq · Y. Zhou
Business School, Huanggang Normal University, No. 146, Xinggang 2nd Road, Huanggang 438000, China

M. A. Shahzad · L. Wang
School of Economics and Trade, Hunan University, Shi Jiachong, Yuelu District, Hunan Province, Changsha City 410079, China

M. Eliw (✉)
Department of Agricultural Economics, Faculty of Agriculture, Al-Azhar University, P.O. Box 71524, Assuit, Egypt
e-mail: moatazeliw@azhar.edu.eg

M. El-Rawy and A. Negm (eds.), *Managed Aquifer Recharge in MENA Countries*, Earth and Environmental Sciences Library,
https://doi.org/10.1007/978-3-031-58764-1_8

1 Introduction

As water scarcity becomes an increasingly urgent issue worldwide, arid and semi-arid countries like Pakistan and Egypt face significant challenges in meeting their freshwater needs [1–4]. Groundwater has long been a vital resource for these nations, supporting agriculture, domestic use, and overall economic growth [5, 6]. However, over-extraction and mismanagement have led to serious problems, including aquifer depletion, saltwater intrusion, and land subsidence [7–9]. In response to these challenges, Managed Aquifer Recharge (MAR) has gained attention as a promising technique for replenishing groundwater resources and ensuring long-term sustainability [10–15].

In this chapter, the objective is to investigate the potential of the MAR in both Pakistan and Egypt by offering a comprehensive overview of their groundwater resources and evaluating the prospects of employing MAR techniques. By examining the similarities and differences between these two nations, we aim to uncover opportunities for collaboration and learning from each other's experiences.

The chapter unfolds in the following manner: Sect. 1 provides a succinct overview of Pakistan's groundwater resources, focusing on the Indus Basin Irrigation System, groundwater's role in agriculture, and the management challenges associated with this crucial resource. Section 2 shifts our attention to Egypt, investigating the condition of its groundwater resources. In Sects. 3 and 4, we explore the potential for MAR in Pakistan and Egypt, respectively, emphasizing the advantages, challenges, and possibilities for implementation. Section 5 consolidates our findings, draws conclusions, and proposes potential pathways for effective MAR management in both countries. Through the comparison and contrast of these two nations, we aim to produce valuable insights that can aid in the improvement of groundwater management practices amid increasing water scarcity.

By examining the potential of MAR within the contexts of Pakistan and Egypt, we strive to offer an in-depth understanding of the current groundwater situation and the part MAR could serve in securing sustainable water resources for future generations. Through this analysis, we aspire to stimulate further research, cooperation, and the adoption of efficient MAR approaches that can mitigate water scarcity in these countries and beyond.

2 Groundwater Resources of Pakistan—A Brief Overview

2.1 *Historical Perspective*

Pakistan has a long history of using groundwater to irrigate agricultural land. The Greek explorers as early as early as 2500 BC documented the existence of horizontal well systems (karezes) in Baluchistan. Before to the invention of surface irrigation, Persian wheels played a significant role in flood-recession agriculture in the Indus plains. However, Pakistan's current level of groundwater exploitation is unparalleled. In Pakistan's most populous province, Punjab, groundwater sources now provide about 40% of farm gate water supplies, up from only 8% in 1960. The development of shallow wells and deep tubewells has contributed significantly to the increase in the area under agriculture in Pakistan during the past few decades, with public and private groundwater extraction responsible for 75% of the increase in water supplies over this 25-year period. Between 1960 and 1985, it also made it possible for cropping intensity to increase by 39% [16]. The unusually flat landscape, inadequate natural drainage, and the semi-arid climate with considerable evaporation gradually led to water logging. By the beginning of the nineteenth century, the Rechna and Chaj Doabs reported the issue for the first time. To assess the scope of the issue and provide remedies, the Punjab government established a waterlogging board in 1912 and a drainage board in 1917. Several corrective actions were undertaken and implemented, including lengthening canal closures, reducing canal supplies, planting eucalyptus groves, lining canals, and building stormwater drains [17]. A 1976–1979 nationwide hydrological survey found that in 22% of the Indus Basin, the water table was within 6 feet of the ground surface, while in another 30%, it was within 10 feet [18]. The government implemented a massive land reclamation initiative known as the Salinity and Control and Reclamation Projects (SCARP) in response to the worrisome growth of areas with high water tables [19]. Vertical drainage was the main strategy used in this effort to combat high groundwater tables. The first significant tubewell drainage project got underway in 1954. A second advantage of the drainage tubewells was immediately noticed. In areas where tubewells were used to pump fresh groundwater, they not only lowered groundwater levels but also increased the amount of water available for surface irrigation [9, 20]. The Punjab Soil Reclamation Act was enacted in 1952. This act established the framework for the Soil Reclamation Board to design and manage drainage tubewells to manage salinity and water logging. Subsequently, the Board was placed on indefinite hiatus, and the Provincial Irrigation and Electricity Department eventually received executive authority from it. However, licensing regulations were initially drafted in 1965 [21]. By 1995, Punjab, Sindh, and—to a lesser extent—Khyber Pakhtunkhwa (KPK) Province had more over 15,000 public deep tubewells. Among these, 12,000 produced fresh groundwater. Due to a lack of effective disposal options in saline groundwater locations, drainage wells were primarily developed in areas with fresh groundwater, even in Sindh Province where there are significant amounts of saline groundwater. By this time, there were about 10,000 tube-wells in KPK province and Balochistan combined. By the late 1980s, groundwater development could now be completely entrusted to the private sector, according to the National Commission on Agriculture's Report. Government funding for the installation of private tubewells largely came to an end [22, 23].

2.2 The Indus Basin Irrigation System

The Indus Basin is Pakistan's largest and is rife with human interference. These operations, which involve constructing dams, canals, and barrages, stem from the massive IBIS. As Pakistan's water needs grow increasingly complex and vital, these operations are becoming more significant (Fig. 1). Since the Indus Basin is shared by Pakistan (47%), India (39%), China (8%) and Afghanistan (6%), the transboundary nature of Pakistan's water resources adds another level of complexity [24]. In a recent report, Young et al. [25] carefully evaluated Pakistan's water resources. They show that Pakistan currently has an average yearly total of 226.4 billion m^3 (BCM) of sustainable water resources. The Kharan desert provides 1.3%, the Indus Basin 96.3%, and the Makran coastal drainage 2.4% to the whole system [24]. Groundwater has emerged as a crucial resource in Pakistan due to the country's seasonal instability, uneven distribution of surface water resources, and difficulty in regulating them [25]. Pakistan now withdraws groundwater at the fourth-largest rate in the world (after China, India, and the USA) and provides residential, industrial, and nearly all of the water used for cultivation [26].

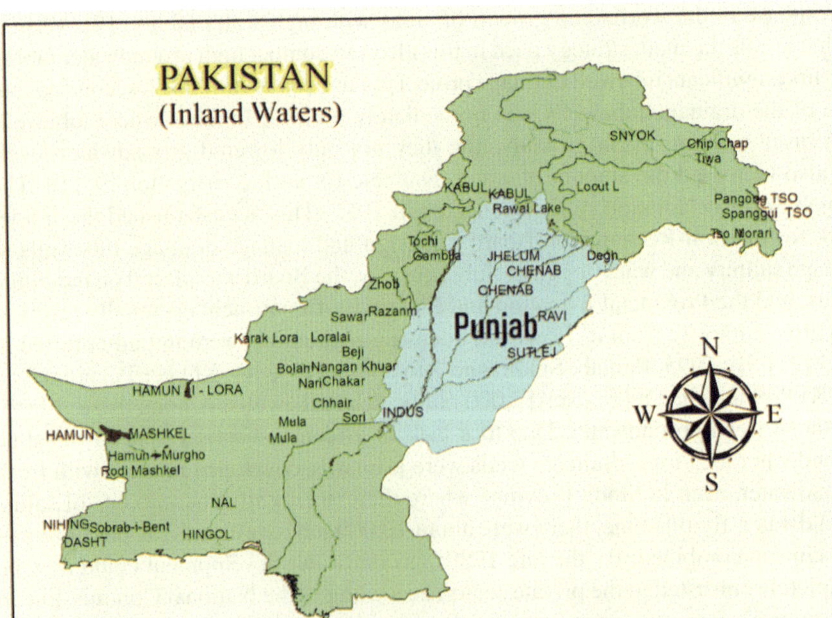

Fig. 1 Inland water resources of Pakistan (Created by QGIS 3.34.1)

2.3 Groundwater as Backbone of Pakistan's Agriculture

According to global figures, agriculture consumes 4,430 km^3 per year of freshwater, accounting for 70% of freshwater extraction [27, 28]. Groundwater contributes significantly; with the world's annual groundwater extraction ranging between 750 and 800 km^3, accounting for one-sixth of all freshwater resources extracted [29]. Currently, groundwater abstraction has gone beyond the rate of recharge, leading to diminishing levels and a potential constraint on future availability.

Because of its distinct advantages, the use of groundwater in agriculture is appealing. These benefits include reliability, timely and on-demand availability, accessibility, and lower per capita investment [25]. The advantages of groundwater outweigh the advantages of surface water volumetric access. If there is surface water accessible, it is preferable not to rely heavily on groundwater extraction in order to slow the aquifers' depletion. As a result, using groundwater as an irrigation resource is a natural option for both small and large farmers looking to improve agricultural productivity and profit margins. This is precisely what has occurred in Pakistan over time.

Irrigated agriculture accounts for about 85% of agricultural production in Pakistan, making it the country's main consumer of water resources [30]. Pakistan has a total land area of 79.61 mha. About 22.15 mha are used for crop production, 4.47 mha are forested, and 25.60 mha are not suitable for cultivation. Approximately 85% of the total cultivated area (18.67 mha) is irrigated using various sources of irrigation, while the remaining 16% is sown under dry farming conditions [31]. Punjab province in Pakistan produces the most grains, livestock, and other agricultural products, accounting for 80% of the country's total cultivated land. Around 82% of Punjab's agricultural land is irrigated, with freshwater from canals and groundwater from tube-wells serving as the primary sources of irrigation [32]. This data demonstrates that Pakistan's agriculture is heavily reliant on irrigation and is extremely vulnerable to changes in the water supply.

2.4 Groundwater Management Challenges in Pakistan

Pakistan's water resources have been poorly managed since the country's independence in 1947. Ineffective water management, combined with rising population, has put enormous strain on water supplies [5], which is reflected in the country's transition from a water-abundant country in the 1950s with per capita water availability of 5260 m^3 to a water-stressed country in 2012 with water availability of 1092 m^3 per person [33, 34]. The nation, which has a population of over 200 million people, is expected to become a water-scarce country by 2030 [25]. This condition has serious implications for the country's agricultural production and food security.

IBIS in Pakistan was built around a century ago to prevent drought and crop failures and to increase settlement opportunities [35]. IBIS is a gravity-based system, which usually requires low management and operational costs. This is a benefit, but it also makes IBIS a rigid system, which its main drawback. IBIS is mainly concerned with the water supply and has little to do with crop water demand. It can only deliver water when there is enough of it. As a result, it is unable to satisfy crop water demands as the seasons change [36]. Another factor limiting Pakistan's agricultural dependency on surface water supplies is the fact that these resources are scarce and could be further exhausted in the future due to climate change. Consequently, there is often a mismatch between IBIS water supply and actual crop water demand [37]. This disparity between crop demand and surface water supply, combined with the delivery system's unreliability and inefficiency, has resulted in the rapid development of groundwater for irrigation.

Farmers in Pakistan have been using groundwater to supplement their irrigation requirements due to the unreliability of surface water provided by the Indus Basin Irrigation System (IBIS). Farmers also diversified crop production methods by using groundwater, significantly increasing agricultural productivity and income. Furthermore, groundwater supply ensures that irrigation water flow is sustained during dry periods [5]. Thus, groundwater has emerged as an extremely important irrigation resource in the country due to the limited and insufficient availability of fresh water and the unique advantages of groundwater. It has played an important role in Pakistan's rural and agricultural and rural development [38] and now regarded as a strategic resource due to the rising demand for agricultural use and other purposes.

However, excessive groundwater usage in agriculture has inevitably resulted in a slew of sustainability concerns. Punjab province has the most private tubewells of any of Pakistan's five provinces [9]. Groundwater is used not only for irrigation in Punjab but also for household drinking and other household purposes. The heavy reliance on groundwater resulted in unsustainable resource development. Another big factor placing a strain on groundwater supplies is the continuous usage of obsolete and inefficient water-use technologies. For instance, Punjab farmers continue using the inefficient flood irrigation method [39].

The flawed policies that allowed for unrestricted extraction of groundwater for agricultural use have now resulted in groundwater level declines, causing many environmental externalities and sustainability issues [38, 40]. For example, in Punjab, continuously falling groundwater levels have forced farmers to dig deeper boreholes and install larger capacity pumps, increasing the cost of irrigation. Furthermore, groundwater quality is continually declining, as shown by elevated chloride, potassium, and magnesium in the water. Approximately 70% of the province's tube wells pump saline-sodic water [41]. Despite these issues, there is no effective policy mechanism in Punjab to limit tube wells' growth.

3 Groundwater Resources of Egypt

Egypt receives 55.5 BCM/year from the Nile Delta [6], which has a total area of around 30,000 km^2 [42], is surrounded by desert [43]. Almost 60% of the population of Egypt resides in the Nile Delta region. Due to the soil's characteristics and the installation of an irrigation system via adjacent canals across the Nile, agriculture predominates in the area. Between Cairo and the Mediterranean Sea is the enormous Nile Delta aquifer (NDA) [44]. Quaternary deposits with a wide range of hydrogeological traits and geographically variable salinity levels produced the NDA [45]. The second source of water is groundwater in the Sinai and Western Desert. In these places, there is an annual abstraction of 0.5 BCM from aquifers. Egypt's coastal regions receive 1.5 BCM/year in rainfall [46]. These deposits show several periods of aggradation and degradation, which were often associated by variations in sea level [47]. Scientific evidence of rising groundwater salinization in NDA has emerged in recent years, mostly as a result of increased groundwater extraction [45] (Table 1).

Groundwater system in Egypt comprises many aquifers, and in the following, we will discuss them in detail [6, 49]:

Table 1 Water budget of Egypt (2017), all sources and allocation/usage [48]

Water supply	Volume (BCM/year)	Demand by sector	Consumption (BCM/year)	Usage/allocation (BCM/year)
Conventional water sources	Drinking (fresh water only)	1.80	10.70	
Nile (High Aswan Dam)	55.5	Industry	5.40	2.40
Deep groundwater	2.45	Agriculture	61.65	61.00
Rainfall and flashfloods	1.30	Drainage to sea	13.5	
Desalination	0.35	Evap. losses	3.00	3.00
Total supply conventional	59.25	Env. balance	0.20	0.20
Unconventional sources	Total consumption	59.25		
Shallow groundwater (Delta)	7.15			
Reuse of agricultural drainage water	13.50			
Total supply non-conventional	21.00			
Total water supply	80.25	Total water demand or allocation	80.25	

(a) Because of the overlying clay layer, the Nile Delta aquifer is regarded as a semi-confined aquifer. Additionally, it is thought to be a shallow aquifer that receives its major replenishment from surplus irrigation water infiltration or from the irrigation network. It is made up of a substantial layer of sand and gravel that is covered in a clay cap with a depth of up to 50 m. About 85% of Egypt's total groundwater abstractions are provided by it [42]. The yearly extraction from the aquifer is about 6.1 BCM [49].
(b) Nubian Sandstone Aquifer System (NSAS). The NSAS is the biggest fossil aquifer system in Africa, according to the International Fund for Agricultural Development (IFAD) [50], with reserves estimated at 259.293 km^3 (372,960 km^3 with post-Nubian). The NSAS, which is shared by Libya, Egypt, Sudan, and Chad, encompasses around 2.2 million km^2, of which 235,000 km^2 (or 11%) are in Chad, 828,000 km^2 (or 38%) are in Egypt, 760,000 km^2 (or 34%) are in Libya, and 376,000 km^2 (or 17%) are in Sudan [51]. The hydraulic head varies from 570 m above sea level west of Darfur to 78 m in the Qatar depression, with the highest depth of NSAS reaching 4,500 m [52, 53]. The fresh layer's thickness varies from 200 m in East Owinat to 3,500 m north of El-Farafra Oasis [49]. A few locations along the eastern edge of the NSAS were found to get recharge from the Nile River due to groundwater input from the Blue Nile and precipitation in a few mountainous regions, according to certain research. Due to leakage into confining beds, discharge into depressions, and evaporation in regions with shallow groundwater tables, low infiltration rates were calculated compared to the natural groundwater flow. It is, therefore, regarded as a nonrenewable groundwater resource, given the state of the climate today [54].
(c) More than half of Egypt is covered by a fissured carbonate rock aquifer that stretches from the Sinai to Libya. It serves as a confining layer on top of the Nubian Sandstone Aquifer System and features several natural springs. There are no accurate estimates of the aquifer recharge or of its potential [49].
(d) Groundwater is channeled toward the Qattara Depression from the Moghra aquifer in the northwest desert. Rainfall and lateral inflow from the Nile aquifer replenish it. Only freshwater is there, and as you move north and west, the salinity rises [49]. Because there is salty water underlying the freshwater lenses, the coastal aquifer "system is refilled by rainfall, and the abstractions are restricted." MWRI [49] replenished by trace amounts of infiltrating precipitation (Fig. 2).

Numerous issues affect Egypt's water sector, including: the Nile River's limited water supply, non-renewable groundwater supplies, rainfall, population growth and demographics, deteriorating water quality, costs of new projects, lack of awareness, the need for greater coordination between various stakeholders, the fact that more than 90% of Egypt's water supply is imported, the effects of international trade agreements, and the effects of globalization on local economies [56–58].

A significant quality risk that restricts groundwater use and reduces the productivity of agricultural regions that depend on irrigation from groundwater wells is salinity [59]. Yihdego and Panda [60] studied the nature and characteristics of salinity with an eye on managing it. Excessive groundwater extraction by humans has a

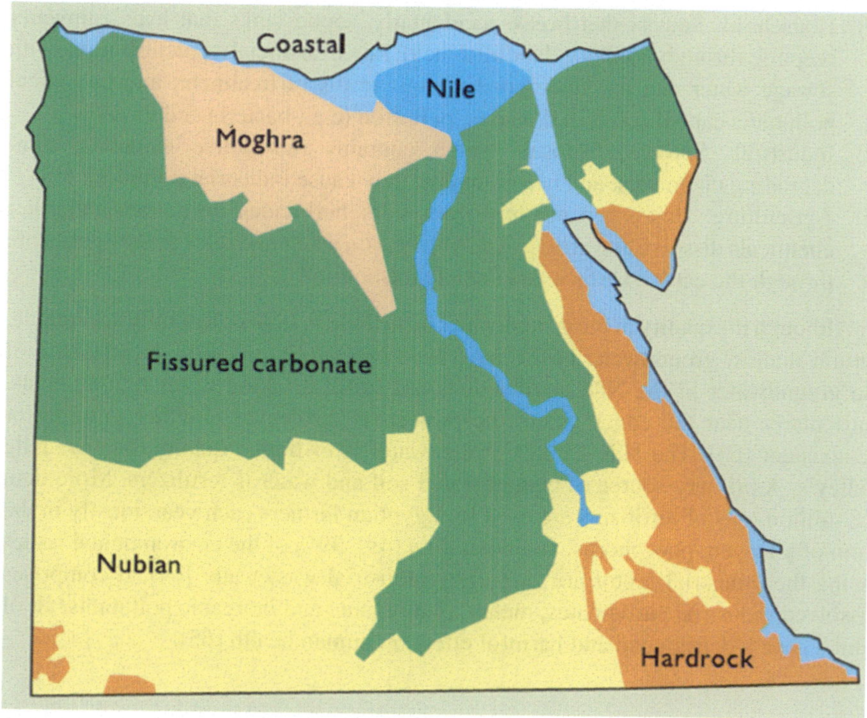

Fig. 2 Major sources of aquifers in Egypt [55]

negative impact on the salinity of groundwater in coastal aquifers like the NDA, and saltwater intrusion (SWI) brought on by sea-level rise (SLR) is also projected [61].

3.1 Groundwater Contamination in Egypt

An essential concern in the creation and administration of water resources is groundwater quality. In reality, the quality issue becomes the limiting factor in the development and management of water resources in many areas of the world due to the rising demand for water in most regions of the world and the intensification of water use. Pollution causes a decline in the quality of both surface and groundwater resources. Aquifer contamination has to get special attention. Four sources are often found to be responsible for groundwater contamination [62]:

(a) Environmental: This sort of contamination results from the features of the environment through which groundwater flows, such as flow-through carbonate rock, intrusion by seawater, and invasion by brackish water from the nearby aquifer.

(b) Household: Sewers that break accidentally, septic tanks that leak, rainwater seeping through sanitary landfills, acid rains, artificial replenishment with sewage water that has undergone various levels of treatment, and biological pollutants can all result in domestic pollution (e.g., bacteria and viruses).
(c) Industrial: Sewage disposal, which contains radioactive elements, non-degrading chemicals, and heavy metals, may cause industrial pollution.
(d) Agriculture: This is because fertilizers, salts, herbicides, pesticides, and other chemicals dissolve and are carried by irrigation water and rains as they permeate through the earth's surface and deplete the aquifer.

Although the quality of the groundwater in the Nile system is generally acceptable, certain shallow groundwater basins have been harmed by pollution. Nearly 20% of the groundwater in the Nile aquifer does not fulfill drinking water requirements, particularly near the edges where the protective clay cover is either minimal or nonexistent [63]. The Nile Delta's groundwater is of higher quality than the Nile Valley's. A primary source of contaminated soil and water is fertilizers. More than 1.8 million tons of fertilizers are used by Egyptian farmers each year, mostly in the form of nitrogen, phosphorus, and potassium [49]. 39% of the environmental issues facing the industrial sector are caused by industrial wastewater [64]. It comprises dissolved industrial particulates, metals, and organic and inorganic pollutants, all of which have a detrimental and harmful effect on human health [65].

4 The Potential for Managed Aquifer Recharge in Pakistan

It is generally known that Pakistan's aquifers are running dry, and on average (1976–2008), 31.5 MAF of excess water flow below the Kotri Barrage without any intentional management for purposes like environmental flows in Dry River reaches predominantly in Punjab [66]. Therefore, it is necessary to quantify the amount of surface water that may be set aside for the purpose without interfering with current usage as well as the amount of groundwater that needs to be recharged, including the capacity of our aquifers to do so. Since fresh groundwater makes up about 70% of Punjab's 21.6 million acres of irrigated land, groundwater storage has already been depleted and irrigation system recharge is already augmenting canal water supply [67]. It is expected that the upper 30 feet of the aquifer can be filled using 50% of the fresh zone, or 7.5 Mac [9]. The potential for water storage in the aforementioned region is 33.75 MAF with a 15% storage coefficient. If correctly managed and integrated with surface water resources, this groundwater storage can function on a cyclical basis for just as long as droughts, which also have a cyclical character. In order to transport water to places with low water supplies, it will be necessary to create new infrastructure or modify existing infrastructure on a priority basis [67].

Prior to the extensive use of irrigation, the Indus Basin's groundwater table ranged in depth from around 40 feet in the Sindh and Bahawalpur regions to about 100 feet in Rechna Doab. The main reason for the rise in water levels was seepage from irrigation systems, some of which had been in use for more than 150 years. According to geologic research, alluvium that is unconsolidated and that has been flooded to within a few feet of the ground surface by seepage practically covers the whole Punjab to depths of 1000 feet or more [68].

The canal system's superposition created additional recharge variables, which caused the water table in and surrounding irrigated regions to increase. According to Greenaman et al. [68], the average groundwater rise rate for Rechna, Chaj, Thal, and central Bari Doabs was 2.2, 2.07, 2.0, and 1.1 ft per year, respectively. The rate of groundwater increase for the Lower Bari Doab has been estimated at 0.7 feet per year. The hydraulic gradients in the vicinity of rivers steadily dropped as seepage from irrigation systems raised the water table, and seepage losses from the rivers also reduced as a result of lower river flows brought on by canal diversions.

The Central Groundwater Board of India created a national groundwater recharge plan to combat the nation's declining water tables. The plan aims to recharge excess runoff totaling about 36.4 billion m^3 (29.5 MAF) in an area of about 450,000 km^2 that has been identified in various regions of the nation experiencing a sharp decline in groundwater levels [69].

The Indus Basin Treaty between India and Pakistan was implemented in 1960, and since then, the situation has completely altered. However, in the case of IBIS, the main obstacle to groundwater recharge is surface water availability. As a result, many river sections are often dry, and the monsoon season is the only time they have surplus flow. The major challenge is managing river flows, especially for groundwater recharge with existing surface reservoirs. Since local recharge from rainfall has already risen owing to water table depletion, the current consensus is that groundwater recharge from rainfall-runoff won't contribute much to the quantity of water in irrigated regions [67].

In regions where the water table is deeper than 60 feet, recharging groundwater by constructing artificial lakes must be a permanent irrigation system element. This so-called command level surface/groundwater storage of flood supply lakes, which may be filled based on the flood share of each canal command, can be constructed during the flood season. The majority of these locations are located at the endpoints of canal commands that fall within the centers of doabs, making it probable that they are far from any potential sources of groundwater replenishment, such as main canals and barrages. Regarding surface water allocations and their effects on increasing the water table in nearby places, this will require appropriate water balance/groundwater modeling studies. It will be necessary to choose areas depending on;

1. The depth of the water table and the distance from sources of recharge, such as rivers and canals.
2. Locations with a deep-water table and strong infiltration rates.
3. Appropriate water allocation during floods based on the severity of flood flows in the affected areas. Moreover, a minimum or additional allotment during each monsoon season.

5 The Potential of Managed Aquifer Recharge in Egypt

Firstly, the potential for managed aquifer recharge in the southern and central regions of Egypt must be considered. In these areas, irrigation water continually recharges the Nile Delta aquifer in the floodplain. As a result, the aquifer serves as a storage reservoir, and it may be used in conjunction with other surface water sources [70]. Three mechanisms contribute to groundwater recharge in the Nile Delta region: (1) rainfall infiltration; (2) downward leakage of surplus irrigation and seepage from the Nile River, irrigation canals, and drains; and (3) groundwater movement between aquifers. The irrigation canals and drains directly bleed water into the Nile Delta aquifer, replenishing it. The downward leakage toward the aquifer in the middle and southern portions of the Nile Delta floodplain ranges from 0.25 to 0.80 mm per day. According to the Research Institute of Groundwater [70], drip irrigation in the desert margins leaks at low rates of 0.1–0.5 mm/day, whereas basin irrigation leaks at high rates of 1–2.5 mm/day [72].

Before exploring the history of dams and engineering structures in the Middle East and North Africa, understanding the significance of Egypt's Nile aquifer system is the key. Dams and engineering structures have a long history in the Middle East and North Africa region. Around 6000 BC, Egypt developed the first irrigation system, using dams and canals to transfer water from the Nile River to arable land. The Marduk Dam on the Tigris near Samarra was built in Iraq (1500 BC), the Marib dam in Yemen (750 BC), and the Falaj (qanat) system that stretched across the Arabian Peninsula, Persia, and Egypt (1300–300 BC). Because Egypt's Nile aquifer system is continuously recharged by return flow and seepage from surface water bodies, groundwater is regarded as the second most important source of water after the Nile River [2, 73]. This aquifer has a total groundwater reserve capacity of 500 billion m^3 (BCM) in the Nile Delta area [74]. Flux from the Nubian Sandstone Aquifer contributes 35% of total Nile Aquifer System recharge. The Nile Aquifer System's (NAS) estimated groundwater abstraction in 2010 was 6200 MCM, which is less than the yield's safe limit of 8400 MCM [75]. Yet, the Nile Delta region's abstraction already surpasses the safe yield [76]. Some areas of the Nile Delta have significantly decreased groundwater recharge, causing seawater intrusion more than 100 km from the shore [77]. The Nubian Sandstone Aquifer System is Egypt's second-largest aquifer (NSAS). The estimated amount of groundwater resources beneath the Western Desert of Egypt is 28,000 BCM, and their natural recharge rate is 120 MCM (i.e., less than 0.03% of the current extraction rate, which is 4000 MCM/year) [78]. This assessment states that the majority of groundwater development in the MENA region is not sustainable. For instance, drawing only 120 MCM (recharge rate/safe yield) from the NSAS in Egypt is not practical or economically viable. MAR, by utilizing all available recharge augmentation strategies, is a more feasible choice. Inducing infiltration from nearby surface water sources and catching water that would have been naturally discharged are further methods. A dynamic steady-state equilibrium known as "sustained yield" can be reached if these sources balance the extraction [79].

Next, an examination of the processes involved in groundwater discharge is necessary. Four processes are involved in groundwater discharge: outflow into the drainage system, direct evaporation, extraction, and groundwater movement between aquifers. The northern parts of the Delta have upward seepage, which discharges groundwater to the drainage system at a daily rate of 0.2–0.9 mm/day [80]. In low-lying places with a shallow groundwater table, there is a possibility of groundwater discharge by evaporation. Groundwater is naturally released by evaporation in lakes, sabkhas, and the Wadi El-Natrun depression. The Nile Delta aquifer, the Moghra aquifer, and the Wadi El-Natrun aquifer all participate in inter-aquifer flow, which is a modest part of the discharge. Between 50 and 106 million m^3/year of groundwater are thought to be moving from the Nile Delta to the Moghra aquifer overall [81]. In the part that follows, groundwater extraction from the Nile Delta aquifer will be discussed.

Transitioning to the topic of abstraction rates, it is crucial to discuss the increasing rate of abstraction from the Nile Delta aquifer over the past 30 years. From 1980 to 2010, there has been a significant rise in abstraction. According to the Research Institute for Groundwater (RIGW) in Egypt, the total annual abstraction rate in 1980 was assessed to be 1.6×10^9 m^3/year, while the net recharge rate to the Quaternary aquifer was projected to be 2.60×10^9 m^3/year. A large rise in abstraction patterns was verified by the RIGW study, which peaked in 1991 at about 2.6×10^9 m^3/year. The annual abstraction totaled 3.5×10^9 m^3/year in 2003 [71, 82]. It got to roughly 4.6×10^9 m^3/year in 2010. The abstraction in Egypt grew roughly linearly by 0.10×10^9 m^3/year, except for 2003 to 2010, when it increased substantially by a rate of 0.20×10^9 m^3/year [45].

In recent studies, researchers have investigated the potential implementation of MAR systems in Egypt. For instance, Soliman et al. [83], the authors investigated the possibility of implementing a MAR system in Sinai, Egypt's Nuweiba alluvial fan aquifer. Groundwater was the primary water source in the past, but a desalination plant has since become the main supply. This plant, however, is prone to breakdowns, potentially leading to severe droughts. The researchers utilized the Geo-Information System–Multi-Criteria Decision Analysis (GIS-MCDA) method to pinpoint suitable locations for MAR systems that would replenish groundwater with flash-flood water. The results revealed that approximately 52% of the study area was appropriate for MAR installation, with 6% demonstrating high potential and 20% displaying moderate potential. The most promising MAR sites were identified in the study area's western part, near the ephemeral stream outlet. This data could support decision-makers in devising sustainable development plans for the area.

Likewise, Hani et al. [84] addressed the growing strain on Egypt's limited water resources in their research. They tackled the growing strain on Egypt's limited water resources, resulting from expanding industrial and agricultural activities, climate change consequences, and rising population. They suggested employing MAR as an innovative approach to augment non-traditional water resources and fulfill essential water needs. The West Delta region was chosen as a case study to assess the feasibility of implementing MAR projects with reclaimed water. The team created a MAR suitability mapping framework using multi-criteria decision analysis (MCDA) and performed a spatially explicit sensitivity analysis to determine the effect of weighted criteria on the ultimate outcomes.

The sensitivity analysis demonstrated that the MCDA-based suitability mapping framework in the West Delta generated reliable results for pinpointing the optimal locations for MAR execution. The results also indicated the potential applicability of the suggested framework to arid regions with comparable data accessibility. Moreover, the research underscored the possible use of suitability maps within sustainable groundwater management strategies to facilitate the practical implementation of MAR initiatives in the West Delta [83].

6 Summary and Conclusions

Pakistan and Egypt are experiencing problems with groundwater management. With its Indus Basin Irrigation System (IBIS), Pakistan boasts the world's largest contiguous irrigation system. IBIS not only irrigates millions of hectares of agricultural land but also holds vast quantities of groundwater, which are vital to the country's water and food supply. Despite a conceptual separation, there is a close interaction between surface and groundwater. Groundwater is present throughout the IBIS primarily because of leakage from its canals. Nevertheless, the stalling of surface storage in the country since 1976 and subsequent silting up of the storage therein over time, have led to increasing demand for groundwater development. Groundwater storage in many of the country's areas is being depleted at a rate of around 1.2 feet per year, which means that it won't last for much longer. By promoting artificial recharge through surface water, which most of the time escapes to the sea, the IBIS must rehabilitate its groundwater potential. The groundwater recharge potential of maintaining river flows in dry reaches for four months is approximately 5.35 MAF. The aquifer has a potential to store 4.78 km3 of flood water. In the southern part of the main Punjab province, the MAR technique is used to increase groundwater. Like Pakistan, Egypt is also using groundwater to meet the growing demand, which is exceeding the safe yield. Surface water supplies about 82% of the Egyptian water demand, while groundwater supplies about 12%. There are many aquifers across the country that have a large hydrogeological potential. Aquifer systems have, however, been overexploited in some places, including the western Nile Delta and the Nile Valley's desert fringes. In addition to demand-side measures, effective supply-side management is imperative to sustain groundwater in Egypt. The Nile Delta region and coastal areas have shown tremendous potential for promoting groundwater recharge.

7 Recommendations

(a) Groundwater Management in Pakistan

1. Develop and implement a comprehensive plan for the management of groundwater under IBIS, addressing the current depletion rates and the conceptual separation of surface and groundwater.
2. Prioritize artificial recharge through surface water to maximize the potential of IBIS groundwater, ensuring that river flows are maintained for optimal recharge.
3. Invest in research and technology for effective MAR techniques, learning from successful implementation in the southern Punjab province.
4. Establish monitoring systems to track groundwater levels, aquifer health, and the success of recharge initiatives.

(b) Egypt's Sustainable Groundwater Strategies

1. Implement strict regulations and sustainable practices to curb overexploitation of aquifers, particularly in critical regions like the western Nile Delta and desert fringes of the Nile Valley.
2. Emphasize a balanced approach to water resource management, giving due consideration to both surface water and groundwater to meet the growing demand.
3. Allocate resources for the development of infrastructure and technologies that support groundwater recharge, especially in high-potential areas like the Nile Delta and coastal regions.
4. Foster collaboration between government agencies, researchers, and local communities to ensure a coordinated effort in achieving sustainable groundwater management goals.

References

1. Abd Ellah RG (2020) Water resources in Egypt and their challenges, Lake Nasser case study. Egypt J Aquat Res 46:1–12
2. Negm AM, El-Rawy M (2022) Sustainability of groundwater in the Nile Valley, Egypt. Springer Nature
3. Qureshi AS, Perry C (2021) Managing water and salt for sustainable agriculture in the Indus Basin of Pakistan. Sustainability 13:5303
4. Qadir A, El-Rawy M (2022) Impact assessment of physiography, subsurface hydraulic gradients and lithologic heterogeneity on the groundwater quality. Iranian J Sci Technol Trans Civil Eng 46:1459–1480
5. Qureshi AS (2015) Improving food security and livelihood resilience through groundwater management in Pakistan. Glob Adv Res J Agric Sci 4:768–710
6. El-Rawy M, Abdalla F, El Alfy M, Abdalla F (2020) Water resources in Egypt. In: Hamimi Z, E-Bamfjfhae RY (eds) The geology of Egypt, Cham https://doi.org/10.1007/978-3-030-15265-9_18
7. El Tahlawi MR, Farrag AA, Ahmed SS (2008) Groundwater of Egypt: "an environmental overview." Environ Geol 55:639–652
8. Razzaq A, Qing P, Abid M, Anwar M, Javed I (2019) Can the informal groundwater markets improve water use efficiency and equity? Evidence from a semi-arid region of Pakistan. Sci Total Environ 666:849–857

9. Razzaq A, Liu H, Xiao M, Mehmood K, Shahzad MA, Zhou Y (2023) Analyzing past and future trends in Pakistan's groundwater irrigation development: Implications for environmental sustainability and food security. Environ Sci Pollut Res 30:35413–35429
10. El-Rawy M, Zlotnik VA, Al-Raggad M, Al-Maktoumi A, Kacimov A, Abdalla O (2016) Conjunctive use of groundwater and surface water resources with aquifer recharge by treated wastewater: evaluation of management scenarios in the Zarqa River Basin, Jordan. Environ Earth Sci 75:1–21. https://doi.org/10.1007/s12665-016-5946-1
11. Al-Maktoumi A, El-Rawy M, Zekri S (2016) Management options for a multipurpose coastal aquifer in Oman. Arab J Geosci 9:636. https://doi.org/10.1007/s12517-016-2661-x
12. El-Rawy M, Al-Maktoumi A, Zekri S, Abdalla O, Al-Abri R (2019) Hydrological and economic feasibility of mitigating a stressed coastal aquifer using managed aquifer recharge: a case study of Jamma aquifer, Oman. J Arid Land 11:148–159. https://doi.org/10.1007/s40333-019-0093-7
13. Al-Maktoumi A, Zekri S, El-Rawy M, Abdalla O, Al-Abri R, Triki CMR, Bazargan-Lari MR (2020) Aquifer storage and recovery, and managed aquifer recharge of reclaimed water for management of coastal aquifers. Desalin Water Treat 176:67–77. https://doi.org/10.5004/dwt.2020.25499
14. Wendt DE, Van Loon AF, Scanlon BR, Hannah DM (2021) Managed aquifer recharge as a drought mitigation strategy in heavily-stressed aquifers. Environ Res Lett 16:014046
15. Amin HM, Gad A, El-Rawy M, Abdelghany UA, Sadeek RA (2022) Improvement of partially treated wastewater quality by soil aquifer treatment in upper Egypt. J Eng Sci Technol 17:0689–0712
16. Sadaf M, Jabbar A (2022) Agent based model for Development of Groundwater Framework vis-à-vis Surface Water Management for improved Returns of Water Resources in Irrigated Agriculture of Pakistan. ESS Open Archive. https://doi.org/10.1002/essoar.10511819.1
17. Rehman G, Jehangir WA, Rehman A, Aslam M, Skogerboe GV (1997) Salinity management alternatives for the Rechna Doab, Punjab, Pakistan: volume one-Principal findings and implications for sustainable irrigated agriculture, IWMI (Accessed Pakistan National Program. xii, 89p. (IWMI Pakistan Report R-021.1/IIMI Pakistan Report R-021.1). https://doi.org/10.3910/2009.452
18. Noor R, Inam A, Zahra SM, Shoaib M, Riaz R, Sarwar A, Asif M, Ahmad S (2022) A methodological framework for modeling sustainability visions: a case study of groundwater management in Faizpur distributary, Pakistan. Agric Water Manag 271:107822
19. Droppers B, Supit I, Leemans R, van Vliet MTH, Ludwig F (2022) Limits to management adaptation for the Indus' irrigated agriculture. Agric For Meteorol 321:108971
20. Razzaq A, Liu H, Zhou Y, Xiao M, Qing P (2022) The competitiveness, bargaining power, and contract choice in agricultural water markets in Pakistan: implications for price discrimination and environmental sustainability. Front Environ Sci 670
21. Sahuquillo A, Cassiraga E, Gómez-Hernández JJ, Andreu J, Pulido-Velazquez M, Pulido-Velazquez D, Álvarez-Villa OD, Estrela T (2022) Management alternatives of aquifer storage, distribution, and simulation in conjunctive use. Water 14:2332
22. Sidhu BS, Sharda R, Singh S (2021) Spatio-temporal assessment of groundwater depletion in Punjab, India. Groundw Sustain Dev 12:100498
23. Van Steenbergen F, Oliemans W (2002) A review of policies in groundwater management in Pakistan 1950–2000. Water Policy 4:323–344
24. Watto MA, Mitchell M, Akhtar T (2021) Pakistan's water resources: overview and challenges. In: Watto Ma MMBS (ed) Water resources of Pakistan, Cham. https://doi.org/10.1007/978-3-030-65679-9_1
25. Young WJ, Anwar A, Bhatti T, Borgomeo E, Davies S, Garthwaite IW, Makin I (2019) Pakistan: getting more from water, World Bank. http://hdl.handle.net/10986/31160
26. Qureshi RH, Ashraf M (2019) Water security issues of agriculture in Pakistan, Pakistan Academy of Sciences (PAS), Islamabad, Pakistan. https://www.paspk.org/wp-content/uploads/2019/06/PAS-Water-Security-Issues.pdf
27. Döll P, Hoffmann-Dobrev H, Portmann FT, Siebert S, Eicker A, Rodell M, Scanlon B (2012) Impact of water withdrawals from groundwater and surface water on continental water storage variations. J Geodyn 59:143–156

28. Kinzelbach W, Bauer P, Siegfried T, Brunner P (2003) Sustainable groundwater management–Problems and scientific tool. Episodes-Newsmag Int Union Geol Sci 26:279–284
29. Shah T, Molden D, Sakthivadivel R, Seckler D (2000) The global groundwater situation: overview of opportunities and challenges. International Water Management Institute, Colombo, Sri Lanka. https://publications.iwmi.org/pdf/H025885.pdf
30. Wudil AH, Ali A, Mushtaq K, Baig SA, Radulescu M, Prus P, Usman M, Vasa L (2023) Water use efficiency and productivity of irrigated rice cultivation in Nigeria: an application of the stochastic frontier approach. Sustainability 15:7824. https://doi.org/10.3390/su15107824
31. Pakistan BOS (2020) Economic Survey of Pakistan 2019–20, Ministry of Finance, Government of Pakistan, Islamabad, Pakistan. www.finance.gov.pk
32. Punjab BOS (2019) Punjab Development Statistics, Bureau of Statistics, Government of Punjab, Lahore. Pakistan. http://www.bos.gop.pk/developmentstat
33. Kamal S, Amir P, Mohtadullah K (2012) Development of integrated river basin management for Indus basin, World Wildlife Fund, Lahore, Pakistan. https://wwfasia.awsassets.panda.org/downloads/irbm_2013.pdf
34. WAPDA (2015) Hydro potential in Pakistan, Water and Power Development Authority (WAPDA), Islamabad, Pakistan. https://nepra.org.pk/Policies/Hydel%20Potential%20in%20Pakistan.pdf
35. Jurriens M, Mollinga PP, Wester P (1996) Scarcity by design: protective irrigation in India and Pakistan (Special report/International Institute for Land Reclamation and Improvement), (Liquid gold paper; No. 1). ILRI. https://edepot.wur.nl/76463
36. Bakshi G, Trivedi S (2011) The Indus equation, Strategic Forsight Group, C-306, Montana, Lokhandwala Complex Andheri West, Mumbai, 400, 053. https://www.strategicforesight.com/publication_pdf/10345110617.pdf
37. Qureshi AS (2020) Groundwater governance in Pakistan: from colossal development to neglected management. Water 12:3017
38. Mustafa D, Qazi MU (2007) Transition from karez to tubewell irrigation: development, modernization, and social capital in Balochistan, Pakistan. World Dev 35:1796–1813. https://doi.org/10.1016/j.worlddev.2007.06.002
39. Razzaq A, Rehman A, Qureshi AH, Javed I, Saqib R, Iqbal MN (2018) An economic analysis of high efficiency irrigation systems in Punjab, Pakistan. Sarhad J Agric 34:818–826. https://doi.org/10.17582/journal.sja/2018/34.4.818.826
40. Ahmed W, Ejaz MS, Memon U, Khair S, Khilji AR, Tarin R, Ahmad F, Qureshi AL, Khan MR, Amin M, Latif M (2021) Groundwater resource management for the Kuchlak Sub-Basin, Balochistan, Pakistan, Report No. 160. https://researchoutput.csu.edu.au/files/181264610/181264309_published_report.pdf
41. Latif M, Pomee MS (2003) Irrigation management turnover: an option for improved utilization of limited water resources in Pakistan. Irrig Drain: J Int Comm Irrig Drain 52:261–272
42. Negm AM, Sakr S, Abd-Elaty I, Abd-Elhamid HF (2019) An overview of groundwater resources in Nile Delta Aquifer. In: Negm AM (ed) Groundwater in the Nile Delta. Springer International Publishing, Cham, pp 3–44
43. EGSA (1997) Egyptian General Survey and Mining, topographical map Cover Nile Delta, Scale 1:2.000.000, Egyptian General Survey and Mining (Publishing Center), Cairo, Egypt. https://shop.geospatial.com/publication/K8NFAEA5521Q4RTW7AMPXAHAA6/Egypt-1-to-2000000-Scale-Geological-Map
44. Farid M (1980). Nile Delta groundwater study. M.Sc. thesis, Faculty of Engineering, Cairo University, Cairo
45. Mabrouk M, Jonoski A, Solomatine D, Uhlenbrook S (2013) A review of seawater intrusion in the Nile Delta groundwater system–the basis for assessing impacts due to climate changes and water resources development. Hydrol Earth Syst Sci Discuss 10:10873–10911
46. El Arabi N (2012) Environmental management of groundwater in Egypt via artificial recharge extending the practice to soil aquifer treatment (SAT). Int J Environ Sustain 1(3). https://pdfs.semanticscholar.org/6544/1e544b82d851a3d85c28aa27d79ae88da97a.pdf

47. Diab M, Dahab K, El Fakharany M (1997) Impacts of the paleohydrological conditions on the groundwater quality in the northern part of Nile Delta. Geol J B 4112:779–795
48. MWRI (2017) Egypt and water—Facts and figures, Ministry of Water Resources and Irrigation, Cairo, Egypt
49. MWRI (2013) Adaptation to climate change in the Nile Delta through integrated coastal zone management, Ministry of Water Resources and Irrigation, Cairo, Egypt. https://www.gefieo.org/sites/default/files/documents/projects/tes/3242-terminal-evaluation.pdf
50. Abu-Zeid K, Abdel-Meguid A (2006) Pioneering action in managing the transboundary Nubian Sandstone groundwater aquifer, Mexico City
51. Cedare (2002) Centre for Environment and Development for the Arab Region and Europe (CEDARE) on activities to support African country parties under UNCCD, Cairo. http://water.cedare.int/cedare.int/files15/File2919.pdf.
52. Sefelnasr A, Gossel W, Wycisk W (2007) GIS-based groundwater flow modeling of the Nubian aquifer system, western desert Egypt. University Halle, Halle. https://www.yumpu.com/en/document/view/51627020/the-nubian-sandstone-aquifer-system
53. El-Rawy M, De Smedt F (2020) Estimation and mapping of the transmissivity of the Nubian sandstone aquifer in the Kharga Oasis, Egypt. Water 12:604. https://doi.org/10.3390/w12020604
54. Mirghani M (2012) Groundwater need assessment Nubian Sandstone Basin, Watertrac, book, Nile IWRM-Net. https://www.springerprofessional.de/en/an-overview-of-groundwater-resources-in-nile-delta-aquifer/16360442
55. MWRI (2005) National Water Resources Plan for Egypt-2017. http://faolex.fao.org/docs/pdf/egy147082.pdf
56. El-Rawy M, Makhloof AA, Hashem MD, Eltarabily MG (2021) Groundwater management of the Quaternary aquifer of the Nile Valley under different recharge and discharge scenarios: a case study of Assiut governorate, Egypt. Ain Shams Eng J 12:2563–2574
57. El-Atfy H (2007) Integrated national water resources plan in Egypt, Ministry of Water Resources and Irrigation Alexandria Governorate. https://documents1.worldbank.org/curated/en/561611468234311417/pdf/341800EGY0whit11public10Action0Plan.pdf
58. El-Rawy M, Fathi H, Abdelrady A, Negm AM (2022) Environmental impacts of treated wastewater contaminates on groundwater quality in the Nile River Valley, Egypt. In: Springer International P (ed) Sustainability of groundwater in the Nile Valley, Egypt, Cham, pp 91–120. https://doi.org/10.1007/978-3-031-12676-5_5
59. Custodio E (2002) Aquifer overexploitation: what does it mean? Hydrogeol J 10:254–277. https://doi.org/10.1007/s10040-002-0188-6
60. Yihdego Y, Panda S (2017) Studies on nature and properties of salinity across globe with a view to its management-a review. Glob J Hum Soc Sci Res 17:31–37
61. Oude Essink GHP, van Baaren ES, de Louw PGB (2010) Effects of climate change on coastal groundwater systems: a modeling study in the Netherlands. Water Resour Res 46. https://doi.org/10.1029/2009WR008719
62. Bear J (2012) Hydraulics of groundwater. Courier Corporation https://www.everand.com/book/271567645/Hydraulics-of-Groundwater
63. Alezabawy AK, Basheer AA, Selim ESI (2021) Delineation and evaluation of the groundwater of fractured limestone Aquifer at East of Al Kurimat Area, Egypt: geophysical and hydrogeochemical constraints. Pure Appl Geophys 178:4425–4459
64. SNC (2010) Egypt's second national communication. Egyptian Environmental Affairs Agency (EEAA-May 2010) under the United Nations framework convention on climate change. https://unfccc.int/resource/docs/natc/egync2.pdf
65. Eea (2004) Environmental status report for 2004. https://www.eea.europa.eu/publications/report_2004_0622_154840
66. Khan A, Awan N (2020) Inter-provincial water conflicts in Pakistan: a critical analysis. J South Asian Middle Eastern Stud 43:42–53

67. Basharat M, Hashmi D (2010) Groundwater management and recharge potential as an alternate to mega surface storages. Communicating Water Quality Challenges and Opportunities, World Water-Day. https://www.researchgate.net/publication/291994461_Groundwater_management_and_recharge_potential_as_an_alternate_to_mega_surface_storages
68. Greenaman DW, Swarzenski WV, Bennett GD (1967) Groundwater Hydrology of the Punjab, West Pakistan with emphasis on problems caused by canal irrigation, West Pakistan Water and Power Development Authority. WASID Bull 6. https://pubs.usgs.gov/wsp/1608h/report.pdf
69. Villholth KG, Sharma BR (2005) Creating synergy between groundwater research and management in South and South East Asia. Roorkee, India
70. RIGW (2002) Nile Delta groundwater modeling, Research Institute for Groundwater, El Qanater El Khayreya. https://www.springerprofessional.de/en/an-overview-of-groundwater-resources-in-nile-delta-aquifer/16360442
71. RIGW (2003) Monitoring of groundwater microbiological activities in the Nile Delta Aquifer. In: A study completed for the National Water Quality and Availability Management project (NAWQAM), Kanater El-Khairia. https://www.researchgate.net/publication/324370442_An_Overview_of_Groundwater_Resources_in_Nile_Delta_Aquifer
72. Environics (2010) EIA study for greeter Cairo metro line number 4 phase one, National Authority of Tunnel, Final report. http://nurhosting.com/NAT/EIAMetroLine4Phase1EnglishFinal.pdf
73. Ebraheem AAM, Senosy MM, Dahab KA (1997) Geoelectrical and hydrogeochemical studies for delineating ground-water contamination due to salt-water intrusion in the northern part of the Nile Delta, Egypt. Groundwater 35:216–222
74. Nofal ER, Fekry AM, Ahmed MH, El-Kharakany MM (2018) Groundwater: extraction versus recharge; vulnerability assessment. Water Sci 32:287–300
75. Sefelnasr A, Sherif M (2014) Impacts of seawater rise on seawater intrusion in the Nile Delta aquifer, Egypt. Groundwater 52:264–276
76. Al-Agha DE, Closas A, Molle F (2015) Survey of groundwater use in the central part of the Nile Delta, Water and salt management in the Nile Delta: Report. https://horizon.documentation.ird.fr/exl-doc/pleins_textes/divers16-02/010066349.pdf
77. Alam S, Borthakur A, Ravi S, Gebremichael M, Mohanty SK (2021) Managed aquifer recharge implementation criteria to achieve water sustainability. Sci Total Environ 768:144992
78. Ebraheem A, Riad S, Wycisk P, Seif El-Nasr A (2002) Simulation of impact of present and future groundwater extraction from the non-replenished Nubian Sandstone Aquifer in Southwest Egypt. Environ Geol 43:188–196
79. Sherif M, Sefelnasr A, Al Rashed M, Alshamsi D, Zaidi FK, Alghafli K, Baig F, Al-Turbak A, Alfaifi H, Loni OA, Ahamed MB (2023) A review of managed aquifer recharge potential in the Middle East and North Africa region with examples from the Kingdom of Saudi Arabia and the United Arab Emirates. Water 15:742
80. Drainage Research I (1989) Land drainage in Egypt, Drainage Research Institute, Cairo. https://books.google.com/books/about/Land_Drainage_in_Egypt.html?id=wdQFVsHLOCUC
81. RIGW (1980) Safe use studies for groundwater reservoirs in the Nile Delta and Upper Egypt, Research Institute for Groundwater, Cairo. https://www.researchgate.net/publication/324370442_An_Overview_of_Groundwater_Resources_in_Nile_Delta_Aquifer
82. Eliw M (2022) An economic analysis of the efficiency of irrigation water use in the production of the most important water-consuming crops in Egypt. J Sustain Agric Sci 48:197. https://doi.org/10.21608/JSAS.2022.110363.1326
83. Soliman K, Sallam OM, Schüth C (2022) Delineating MAR sites using GIS-MCDA for Nuweiba Alluvial Fan Aquifer, Sinai, Egypt. Water 14:475. https://doi.org/10.3390/w14030475
84. Hani HM, Nour El Din MM, Khalifa A, Elalfy E (2023) Sensitivity analysis for multi-criteria decision analysis framework for site selection of aquifer recharge with reclaimed water. Sustainability 15:5399. https://doi.org/10.3390/su15065399

Potential Environmental Risks of Aquifer Recharge Using Treated Wastewater in MENA Countries

Heba Fathi, Mustafa El-Rawy, Naglaa Zanaty, and Kamel Mansour

Abstract In several MENA countries, turning to mechanical treatment using activated sludge and secondary and tertiary aeration ponds has significantly enhanced the sewage's both quantity and quality. Using treated wastewater may increase an aquifer's groundwater supply and quality; these options are practical when access to conventional fresh raw water sources gets severely limited. Artificial recharge techniques have previously been identified to boost groundwater levels and improve water quality in several groundwater aquifers. The results showed that treating wastewater before injecting it back into groundwater aquifers is a helpful method for artificial recharge, the findings of managed aquifer recharge (MAR) in the coastal aquifer north of Oman showed that MAR methods combined with managing agricultural wells and relocating public wells may increase the volume abstracted for drinking by 200%. Additionally, the Zarqa River in Jordan was studied, and the findings showed that the water table rose by 29 m after the injection of the treated wastewater. It is recommended to acquire recharged water of adequate quality at a distance of 1500 m from the infiltration basin. However, further research with well-designed monitoring systems are required to assess how the recharge process affects groundwater quantity and quality.

H. Fathi
College of design and architecture, Jazan University, Jazan 45142, Saudi Arabia
e-mail: hamansour@jazanu.edu.sa

M. El-Rawy (✉)
Civil Engineering Department, Faculty of Engineering, Minia University, Minia 61111, Egypt
e-mail: mustafa.elrawy@mu.edu.eg

Civil Engineering Department, College of Engineering, Shaqra University, Dawadmi 11911, Saudi Arabia

N. Zanaty
Environmental Studies Department, National Authority for Remote Sensing and Space Sciences (NARSS), Cairo, Egypt
e-mail: Naglaa.zanaty@narss.sci.eg

K. Mansour
Urban Planning Department, National Authority for Remote Sensing and Space Sciences (NARSS), Cairo, Egypt

M. El-Rawy and A. Negm (eds.), *Managed Aquifer Recharge in MENA Countries*, Earth and Environmental Sciences Library,
https://doi.org/10.1007/978-3-031-58764-1_9

Keywords Managed aquifer recharge · Treated wastewater · Reused wastewater · Water quality · MENA

1 Introduction

Population growth in MENA is 1.6%, higher than the global average of 1.1%, while more than 70% of the limited freshwater supply is used for agricultural use; these rates worsen the water demand-supply imbalance [1]. As a result, MENA countries typically rely on aquifers, which lead to over-exploitation and quality degradation, as well as non-conventional water sources such as treated wastewater, urban runoff, and desalinated seawater to meet their water demands [2]. For many countries in the MENA area, several factors affect how much water is available. The amount of fresh water per person varies from extremely as 200 m^3/year/capita in Jordan to countries with less water stress, such Iraq with 4,340 m^3/year/capita [3]. The extremes-arid climate (high temperatures and significant evaporation for the majority of the year) is also cause of the water issue in MENA.

Dealing with the current water crisis in MENA needs collaborative effort for controlling the resources that nations share and to alter human habits and living requirements [4]. Water scarcity solutions in MENA confront several challenges, including social disapproval (such as water reuse) and high water treatment costs. Additionally, most of the region has no water economic laws and regulations, meaning that water prices are set at the national level and costs are covered independently of usage or taking into account the resources present [5]. Identifying and developing new, alternate water sources is a more interesting and practical solution. Reuse and desalination are frequently the only options left for increasing supply in most MENA countries because the conventional sources of groundwater and surface water have long since been exhausted. The high expense of desalination and the promise of better wastewater management and treatment account for a large portion of the appeal of wastewater reuse [3, 6]. The GCC almost entirely on desalinated water, for instance, the UAE has the biggest desalinated saltwater facility in the world, other MENA poor countries without access to abundant energy resources, such Yemen and Libya, may not be able for desalination since it is energy-intensive; nevertheless, using other renewable water sources should allow [1].

As Sherif et al. [7] emphasized, the majority of dams in the MENA area were constructed to improve groundwater recharge. The findings of this investigation, the recharge effectiveness of dams takes place from 15 to 47%, as a result, MAR technology must be used to enhance the operation of dams. Under the existing political economy of water supply, wastewater reuse supports the belief that recycled water is low quality and therefore must be given to consumers for free or at a very cheap cost; this increases scarcity because there is no opportunity for recycled water.

Reusing wastewater has a lot of advantages. Reduced water resource contamination is one of the environmental advantages besides controlling saline water intrusion through groundwater recharge [8]. Furthermore, reusing wastewater offers irrigation with nutrient-rich water and lessens the need for artificial fertilizers [9]. Additional job opportunities and exportable products are socioeconomic advantages. The danger to public health posed by pathogens, mineral traces, and organic traces in wastewater is eliminated by wastewater reuse [10]. In the MENA region, domestic and industrial sectors generate 13 billion m^3 of wastewater each year, with 6 billion m^3 being handled for reuse.This alternate source could be regarded since the quantity of treated wastewater created is particularly appropriate to the amount of supplied water [11]. Two management approaches allow for significantly more wastewater reuse in the MENA region. In the first instance, recycled water is combined with fresh water sources in an indirect reuse scenario. The second occurs when recycled water is pumped directly through different systems [5]. The national water plan for Egypt aims to use 2400 million m^3 of water per year for agriculture, whereas Abu Dhabi plans for the reuse of all treated water through 2030 [12].

When the aquifer suffers significant drops due to groundwater extraction, there is an imbalance between pumping groundwater and natural recharging. While addressing the region's water needs, reducing present-day extracting and supplying artificial recharge in certain areas of the aquifer would stabilize and stop the fall in groundwater levels [13]. However, one of the most important methods for ensuring the sustainability of groundwater aquifers is artificial recharge [14]. The main goals of aquifer recharge (AR) projects in the MENA area are long-term aquifer conservation, recovery, and emergency storage requirements. Using numerical modeling, El-Rawy et al. [15] and Al-Maktoumi et al. [16, 17] investigated the viability of Managed Aquifer Recharge (MAR) in alleviating a deteriorating coastal aquifer north of Oman. The results demonstrated that controlling irrigation wells and moving public wells in combination with MAR practices could raise the abstracted amount for consumption by 200%. Otherwise, Bekele et al. [18] indicated that the recharge water may polish or improve the aquifer's quality, but the prompting of microbial can also cause a decline in water quality. Al-Maktoumi et al. [19] utilized the MAR as a reliable and sustainable water source in the Duqm district of Oman. Djellali et al. [20] employed a 3D numerical model based on a limited variation approach to simulate groundwater flow from 2021 to 2031 in northeastern Algeria. Increased piezometric levels at the suggested sites during the modeling process using artificial recharge showed a favorable reaction to artificial recharge, indicating a positive effect on the aquifer. In Egypt, Abdelrady et al. [21] used numerical modeling and field measurements to investigate the effectiveness of bank filtration for water supply performance.

Rock-water contacts during aquifer storage during acid rain can increase the levels of arsenic, manganese, iron, and a variety of trace ions in groundwater, based on the pH, organic matter, mineralogy, and microbial activity in the aquifer [22].

Global issues that AR may easily solve include water scarcity, dry climate change, seawater intrusion into coastal aquifers, and depletion of ground water and quality improvement. While avoiding some of the issues of above-ground storage, AR generally needs a significantly smaller area footprint than similar surface water storage in metropolitan settings where land is expensive [23]. Through AR, more water may be made available for irrigation of urban green areas, reducing the impacts of urban heat islands and raising amenity value. When compared to other wastewater disposal techniques, AR is comparatively inexpensive, especially when nutrient removal expenses are taken into account to satisfy environmental discharge criteria to freshwater or marine ecosystems [18, 24].

This chapter discusses the major challenges facing the MENA countries behind sustainable wastewater utilization, and trace organic pollutants are attenuated naturally during aquifer recharge. Finally, recommendations pertinent to future research on aquifer recharge using treated wastewater technologies are suggested to enhance their implementation as advanced sustainable water sources.

2 Recycling Methods of Treated Wastewater

It is common knowledge that water is one of the elements that sustains life on Earth and is a vital component in the food chain. Figure 1 shows the annual water demand and supply measured and expected for 2050 in MENA countries. Unluckily, a number of household, industrial, and agricultural practices have the potential to contaminate this resource. In both urban and rural locations, wastewater treatment is a necessary but frequently problematic process. Since this effluent is frequently released into the environment, particularly the sea, streams, and water bodies, it needs to be cleaned. Numerous microorganisms found in wastewater can seriously infect people, animals, and plants [25, 26].

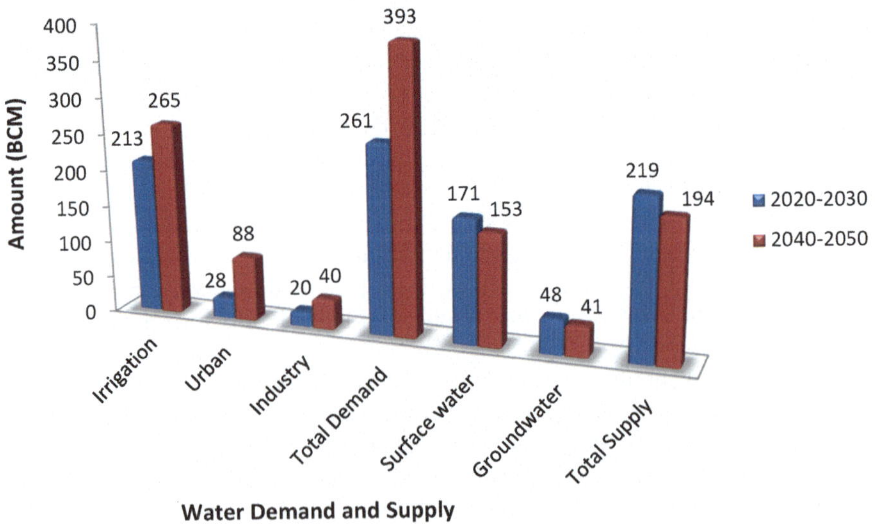

Fig. 1 MENA annual water demand and supply; 2000–2050 BCM. Modified after [2]

Since wastewater is frequently released into the environment, particularly the waterways, streams, and lakes, it must be cleaned. Many different types of microorganisms found in wastewater can lead to significant illnesses for humans, animals, and plants. However, the majority of these microorganisms are benign and can be a valuable resource for bioremediation of stubborn toxic chemicals and inactivation of waterborne pathogens. Therefore, to raise the hygienic quality of wastewater, appropriate physicochemical and biological (secondary treatment processes) treatments should be taken into account [27].

Wastewater may be cleaned up using a variety of techniques, based on the kind and level of pollution. Wastewater treatment plants handle the majority of wastewater by using physical, chemical, and biological treatment techniques. Reuse may indicate that the treatment procedure has well progressed. Due to the stink, the annoyance, and societal views, using wastewater may not be socially desired, but it may be accepted as long as there are advantageous applications [28]. The ways that different countries handle wastewater also differ. In this regard, less than 10% of collected wastewater in Iran, Lebanon, Morocco, and Libya is treated, compared to all collected wastewater in Bahrain, which is treated using tertiary treatment and activated sludge [2].

Mostafa et al. [29] confirmed that Treated wastewater is an essential resource that offers several financial benefits. It is easily accessible near towns and cities where there is a high demand for both fuel and nutrients. Irrigating crops with nutrients from treated wastewater might reduce or eliminate the demand for synthetic fertilizers.

Department of Environmental Protection [30] in Pennsylvania University summarized the benefits of water reuse as follows:

- Reduces demands for surface water or groundwater resources.
- Support reduces the pollutant loading to surface water.
- Respite costly investment for the development of new water sources.
- Economic benefit for the country.
- Save especially useful discharge alternatives.
- Reservation of watersheds.
- Reduces demands on useful water supplies used for drinking water and irrigation.

Heavy metals such as manganese, iron, copper, and zinc are more prevalent in wastewater than in any artificial fertilizer, posing dangers to human health and surrounding ecosystems due to their microbiological and chemical components. Excreta-related infections and disease drivers such as microbes, viruses, and parasites are common in wastewater, posing health concerns to unprotected populations. Unfortunately, numerous microorganisms have demonstrated resistance to biological wastewater treatment methods, so there have been major advances in measuring natural transformation and removal rates [31].

Throughout the MENA region, the removal efficiency of several WWTPs was shown to be negative in case of overload [5]. Effluent treatment may be unsuccessful due to issues with treatment techniques and infrastructure capabilities caused by variable quality and increased effluent volume. For instance, 57% of the wastewater from MENA countries is only partially or never treated [11]. Municipalities used conventional "waste stabilization ponds" to treat wastewater and other municipal wastewater since they lacked the necessary financial resources. Because of the high amounts of chloride and nitrate, the effluent from these facilities is occasionally combined with freshwater supplies and utilized for irrigation, causing major contamination issues. The overflow of stabilization ponds also causes untreated water to contaminate the groundwater supplies and downstream regions [32].

However, in certain MENA countries, switching to mechanical treatment with "activated sludge" with tertiary treatment has significantly enhanced the quality and amount of treated wastewater. Nevertheless, the anticipated growth in MENA population would inevitably produce more wastewater [1]. Figure 2 shows the amount of wastewater that is treated and used again in MENA countries. Discharging treated wastewater is a situation that makes it possible for any residues of chemicals to reach the environment, especially if they are still in use and not subject to any controls or regulations. In many MENA countries, this treated wastewater is utilized for a variety of purposes, including irrigation of gardens and agricultural land, groundwater recharge, and even indirect potable reuse. Jordan and Tunisia are regarded as leaders in treated wastewater reuse. The most commonly used method involves discharging wastewater into surface water, diluting it, and then using it for irrigation of land along the Nile and in the Nile delta [5]. Two significant barriers to intended reuse applications exist in the region: a shortage of (i) sewage collection and (ii) treatment. As a result, a lack of any of the above parts reduces recycling percentages. The barriers to wastewater collection and treatment are mainly economic in nature. In numerous MENA nations, majority of the population in congested metropolitan areas continues to use septic tanks or other on-site sanitation methods [33]. The following criteria determine the effective consequences of wastewater reuse [24]:

1. The difference between the cost and quantity of recycled and traditional water.

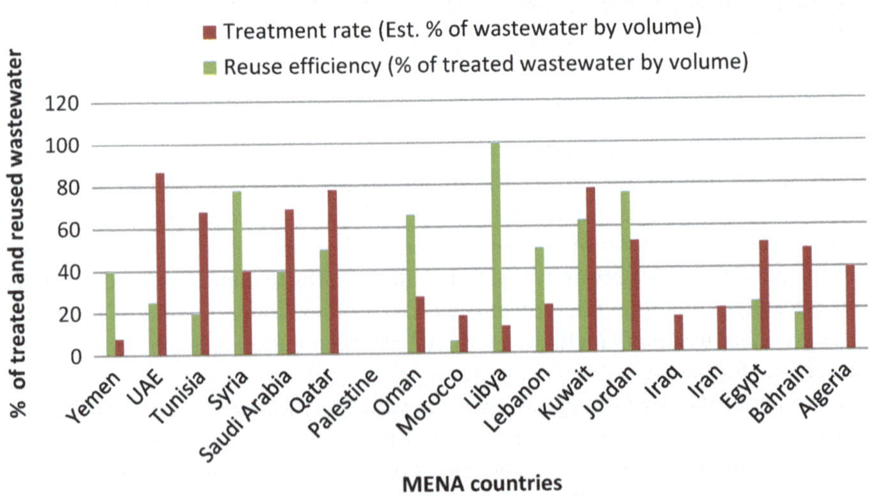

Fig. 2 Percentage of treated and reused wastewater in MENA. Modified after [24]

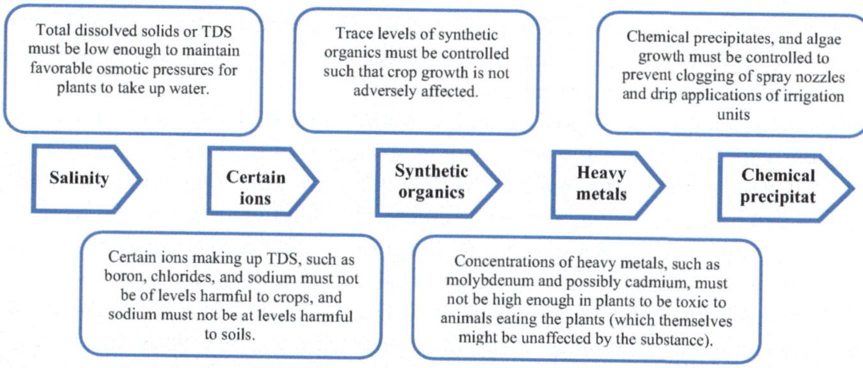

Fig. 3 The main requirements for treated wastewater. Modified after [28]

2. The degree to which these subsidies lead to an inefficient distribution of water supplies, or the financial flexibility of demand.
3. How much does the scarcity value of water, the expense of substitutes, or the preference for low-value applications restrict high-value water usage?
4. The amount that can be reused can encourage development in undersupplied sewage treatment areas.

It is possible that reuse in the region may be constrained by causes other than a lack of wastewater transportation and treatment given the low incidence of reuse in so many MENA countries with water crisis difficulties such as Bahrain, Libya, Egypt, Morocco, and Tunisia [34]. This might be as a result of the product of reused wastewater being less than the delivery cost. Additionally, users may be more aware of actual or perceived quality variations between conventional and recycled water. Given that facilities for distributing recycled water involve investments in expensive civil works like irrigation canals or piped water systems, financial obstacles may also be a factor [35]. Since the most use of treated wastewater is irrigation, Choukr-Allah [28] summarized in Fig. 3 the most important criteria to be taken into account to obtain environmentally suitable water resources for plants and people.

3 History of Reusing Treated Wastewater in Aquifer Recharge in MENA Countries

Governments, decision-makers, and planners in the MENA region are becoming more interested in utilizing unconventional water resources, such as recycled wastewater, to fulfill water demand as a result of the region's rising water shortage [11]. Wastewater reclamation can be used to improve water quality, protect the environment and delay expensive investments in water storage and desalination [36].

More places are expected to find enhanced wastewater management appealing as MENA countries continue to expand economically, as indicated by recent attempts to improve wastewater management in Morocco [24].

The consequences of treated wastewater flow into Jordan's Zarqa River and the underlying unconfined limestone Hummar Aquifer were examined [37]. They constructed an aquifer conceptual model to better grasp water dynamics. The study reported that modeled conjunctive usage and MAR applying mixed treated wastewater with groundwater give a significant case study of advanced water management techniques in the MENA area, as the results indicated that the water table rose by 29 m following the inception of the treated wastewater. Amin et al. [38] discussed an appealing method utilized in Egypt to safely reuse wastewater in unrestricted irrigation. The study also revealed that 1500 meters from the infiltration basin is the ideal distance for obtaining recharged water of acceptable quality.

Recently, Ajjur and Baalousha [6] indicated that eight MENA countries is using handled wastewater discharge into a water resource in AR projects, and 13 MENA countries used rainfall, also they represented the percentage of aquifer recharge sources as 41%, while the treated sewage reached 25%. If the recharged water were of poor quality, that may pose health risks. Let's think about the utilization of several such aquifers for drinking water. The problem gets much more serious and needs immediate protective action to save the aquifers. Choukr-Allah [28] identified the environmental problems that are expected to occur with continuous aquifer recharge using treated wastewater

(1) Potential chemical contamination of the groundwater.
(2) Wastewater not properly treated can create potential public health problems.
(3) High amounts of suspended particles in treated wastewater might block irrigation nozzles and soil capillaries.
(4) Some soluble components in wastewater may have harmful amounts for crops.
(5) Treated wastewater is available year-round, but irrigation demand varies seasonally.

As a result, MENA nations should investigate a high-quality source of water that fulfills WHO quality requirements. They must additionally manage aquifer recharge (MAR) technologies and check the quality of recharged water to avoid health risks [6]. There are still chances to improve the state of the art in AR. To further improve the quality of groundwater before recovery, these consist of thoroughly recorded field experiments with groundwater additives or accessible reactive barriers that alter the biogeochemical ecosystem. Additionally, there are still prospects for increased scientific regulatory adoption and a standardized method to evaluate natural treatment so that an aquifer is included in a treatment train, particularly in drinking water systems. In order to plan and model the effects of AR, pollutant emission and transfer registry data may also be used more often. To enable advantageous geochemical reactions or dilution, source water chemistry data may be specifically correlated with scattered or concentrated pollution in aquifers [18]. Table 1 shows the most important laboratory and in-situ applications of aquifer recharge using treated wastewater in some MENA countries.

Table 1 List of applications for treating wastewater that recharges aquifers

Study	Main outcomes
Hassan et al. [14]	A groundwater modeling system was used to predict the impact of artificial recharge on the volume of groundwater in the Dibdibba unconfined aquifer in Iraq. The findings show that, at 5000 and 10,000 m^3/day pumping rates, respectively, the injection of treated water through 20 wells increased the water table in more than 91 and 136 km^2
Rahman et al. [9]	MAR methodology was used to analyze a number of water resources management methods employing treated wastewater in the Northern Gaza Strip. The prospective effects of these approaches on groundwater resources, farming, public health, and society as a whole are positive
Missimer et al. [39]	Following an evaluation of the comparative costs of treated residential wastewater in Saudi Arabia for both indirect and irrigation use, ARR storage, and treatment, In summary, one important way to reduce the cost of providing clean drinking and irrigation water to rural communities in dry environments is to employ MAR for the storage and treatment of highly polluted wastewater
Bouri and Dhia [40]	The Teboulba aquifer in Tunisia now has better geochemical and piezometric characteristics thanks to the artificial recharge efforts done since 1971. A useful strategy for controlling water resources is artificial recharge using dam water. Installations required for recharge operations are straightforward and reasonably priced. The recharge via direct injection into the aquifer is the most effective approach for building freshwater barriers against saltwater intrusion

4 Characterization of Wastewater Effluent for Managed Aquifer Recharge

Artificial recharge provides an effective form of water storage, allowing for improved resource management, and can also impact water quality indices. The use of treated wastewater is an essential component of water management planning, notably for the artificial replenishment of groundwater resources and agriculture [14].

As a result, recovered wastewater is utilized to increase the amount and quality of groundwater in an aquifer. Several studies [37, 39] have shown that such alternatives are viable when traditional fresh raw water supplies become extremely restricted. Artificial recharge techniques have previously been identified to boost groundwater levels and improve water quality in several groundwater aquifers [39].

Due to the enormous variety of chemicals that might possibly be present in wastewater, researchers have created a number of water quality standards that recycled water must achieve before being recharged to safeguard human health [37, 39]. The poor removal during treatment may be caused by:

- Treatment conditions (such as climatic factors or the treatment plant's age) may impact on the efficiency of WWTPs.
- The operating requirements (such as lower retention periods while the treatment facility is overloaded and surplus flows that overwhelm the plant).

- The kind of polluted influent wastewater that may reduce the efficiency of biological treatment, rearrangement, deterioration, and/or accumulation of contaminants.

Qadir et al. [41] summarized the obstacles facing the process of reusing wastewater in the MENA countries specifically, as shown in Fig. 4.

Ajjur and Baalousha [6] found that the following elements should be taken into account while choosing AR sites:

- Aquifer features, land use/cover, rainfall characteristics, evaporation, transpiration, drought indicator, runoff statistics, sediment, etc.)
- Factors of socioeconomic status (such as demography & future population projections, implementation cost, and operation & maintenance costs).

Bekele et al. [18] concluded that clogging became the main important operating issues in AR since it limits the amount of water that can be recharged, raising the actual unit cost of stored water. Clogging occurs over time as a result of the interaction of the source water (including its contents), native groundwater, and porous material. This can lower the porousness of the nearby aquifer, well zone, or infiltration region. Clogging-induced permeability decreases result in a decrease in injection rate and/or an increase in hydraulic head.

Assessing the biological, chemical, and physical characteristics of groundwater as well as the cause of any potential pollution that could be causing the quality of the water to deteriorate is the process of determining water quality. Several indices

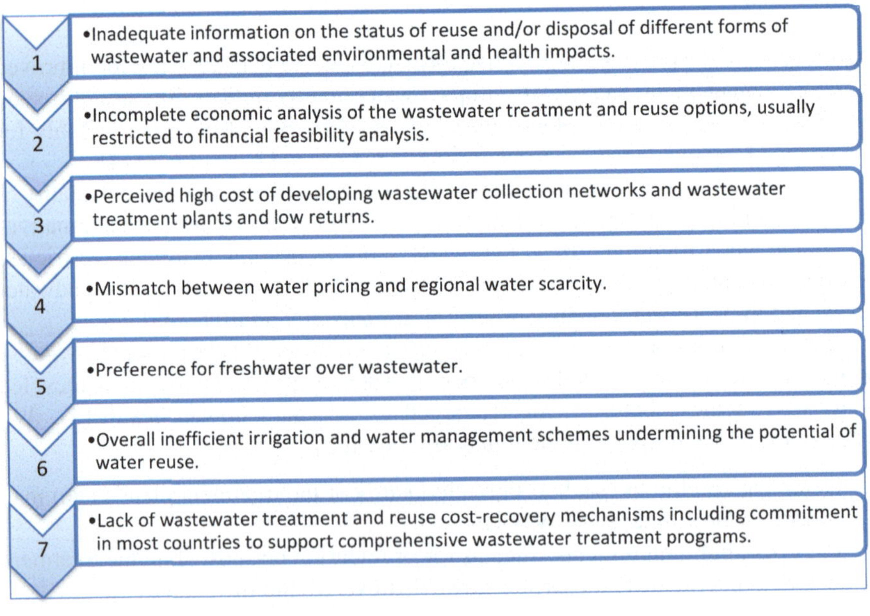

Fig. 4 The main obstacles of wastewater reuse in MENA countries. Modified after [41]

of water quality can be divided into: (1) Chemical: pH, DO, BOD, nutrients (such as phosphorus and nitrogen), organic and inorganic elements, and toxicants; (2) Biological:algae and microbes; (3) Physical: temperature, turbidity, salinity, color, suspended solids, and dissolved solids; and (4) Appearance: odors and floating substances [42, 43].

The consequences of groundwater quality parameters generally are compared with the permissible level for a specific purpose in the identification of contaminated sources and their extension [44, 45]. Their solubility will control heavy metal mobility, both in the recharge water and within the aquifer after mixing of treated water with groundwater [18]. Insoluble metals can be eliminated by filtration of a porous medium, which can lead to the physical blockage of an infiltration basin or injection well. Fluid speciation followed by dissolution or adsorption inclination to surfaces that are accessible, such as metallic oxides, organic matter, and mineral substances, can regulate soluble species concentrations [24].

5 Impacts of Treated Wastewater Reuse on Aquifer Recharge

Managed aquifer recharge (MAR) is one of the techniques which integrates surface and groundwater. MAR uses the excess runoff or reused urban wastewater to recharge aquifers and substitutes the recharge deficiency resulting from reservoir construction or overdraft of groundwater. Water quality would be affected, as a result of this process. The degradation of water quality can be caused by natural conditions or anthropogenic factors; human exploitation of groundwater can lead to deterioration. Aquifer particles may be subjected to air during extraction or recharge operations, which can alter the direction of groundwater flow and enable naturally occurring toxins like fluoride, salt, arsenic, and radium to accumulate in the pure water. Low water quality and related health effects may arise from this. However, pollutants that are emitted from farms and factories, both pharmacological and biological, can also cause a decline in the quality of water [46]. In addition, facilities have been created with the intention of increasing aquifer recharge through the use of processed wastewater or water moved from rivers during periods of peak flows. The artificial recharge with wastewater depends on the exchange effects of soil and water on each other. Thus, soil plays a crucial role in the process of purifying the recharged wastewater [47, 48].

In Egypt, the International Commission on Irrigation and Drainage (ICID) environmental Checklist has been utilized to evaluate the environmental impacts of using irrigation wastewater on groundwater quality and quantity in El-Gabal El-Asfar farm. The investigation identified ways to reduce groundwater contamination and other effects of using wastewater effluent for irrigation [49].

As Egypt is facing a shortage in water resources availability as a result of population growth and food requirements. This was examined in various studies,

El-Fakharany [48] investigated and quantified the artificial recharge of treated wastewater in Abu Rawash through the application of the Soil Aquifer Treatment (SAT) system in order to restore treated wastewater under the prevalent conditions. In addition, this study applied an environmental impact assessment (EIA) study before and after the experiment's operation. According to the author, reusing non-conventional water resources increases the aquifer potentiality for low-productive aquifers, including those in the Abu Rawash area, encouraging recovery of the over-exploited aquifer and reducing groundwater depth. Aquifers with deep groundwater are more convenient to realize SAT because of the great thickness of the unsaturated zone. Soil has a significant role in the process of purifying recharged wastewater [48]. BOD and COD concentrations declined by 50–80%, which enhances the recharge efficiency and completion of treatment through soil stratification. According to the EIA study, after the experiment was run, the positive benefits increased from 15 to 64% whereas the negative effects fell from 66 to 9%.

Figure 5 shows the percentage of negative and positive impacts: (A) before recharge, (B) after recharge. The feasibility of artificial recharge with wastewater depends largely on the quality of wastewater and the soil's capacity to enhance its quality. The findings showed that treating wastewater and utilizing it as a means of artificial recharge for groundwater aquifers is a viable strategy. Further study is needed, though. In order to assess how the recharge technique affects the amount and quality of groundwater, an efficient monitoring framework is also required.

Moreover, Ajjur & Baalousha [6] reviewed the challenges of MAR in the Middle East and North Africa (MENA) region based on a survey of 142 previous research studies. Their survey revealed the applied methods of MAR, in addition to showing the technical and socio-economic constraints that lead to MAR failure in MENA countries. This article presents a framework to assess the MAR feasibility in the region that is suffering from climate change impacts, in addition to provide recommendations for MAR future research and designs in the MENA region.

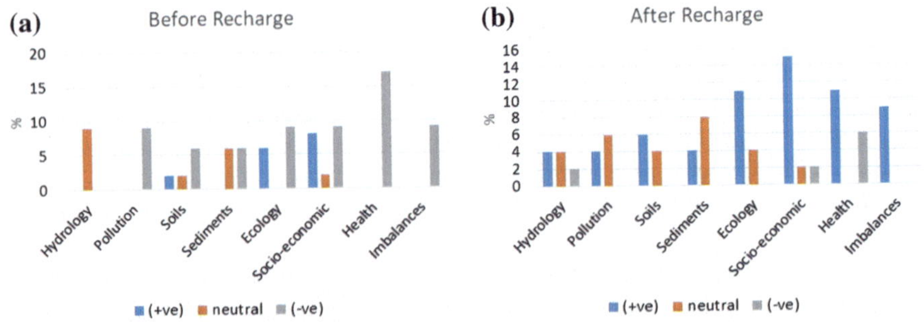

Fig. 5 Percentage of negative and positive impacts: **a** before recharge, **b** after recharge. Modified after [48]

6 Socio-economic, Health, and Environmental Risks of Aquifer Recharge

According to the quality and the efficiency of treated recharge water, it may include varying concentrations of hazardous microbes and harmful substances, as well as minerals and heavy metals [50]. Using various sources of water with different qualities for MAR systems; specially treated wastewater might cause high risks to public health. In the case of using the treated wastewater for MAR recharge, the intricacy of reacting transport processes inside the unsaturated region reveals two primary limitations which must be taken into consideration.

6.1 Environmental Impacts

Wastewater reuse is another method to reduce the amount of wastewater that is released into collecting waterways and conserves more freshwater for the environment. Reusing wastewater can serve several purposes, including but not limited the agricultural sector, manufacturing, landscape watering and urban, replenishment of groundwater, and development of wetlands. Reusing wastewater is largely motivated by the environmental benefits it offers, but socioeconomic considerations are also quite important [51]. However, reusing wastewater has a risk of being harmful to the environment, so caution should be exercised when rushing for many advantages. Finding the right balance to produce an environmentally positive income is the difficult part [52].

In the majority of the world, anthropogenic impacts on freshwater systems are noticeable. Excessive extraction, mostly for agricultural purposes, has seriously deteriorated aquifers, streams, and lakes. Reducing the impact of human activity on the environment by replacing wastewater for non-potable purposes has been commonly recommended [53].

6.2 Effects of MAR Recharge on Water Quality

(A) **Trace Metals**

Metals pose an issue in recharge water, which concerns using treated wastewater at the beginning since the level of several trace metals are fairly uncommon in almost all waterways. Various research has revealed that although the amounts of trace metals in drainage and surface water can vary significantly, they are often below permissible limits, with the exception of lead and iron [54, 55]. Iron and lead present relatively lower risks to health. Recharge water from treated wastewater may include quantities of metallic substances, which Iron, zinc, copper, and lead are the most frequent metals. Other trace metals such as manganese, aluminum, chrome, arsenic, selenium, mercury, cadmium, molybdenum, nickel, etc. from varied sources can also be available. They are produced from different origins including mostly products that could be consumed by humans, material corrosion in the water transmission and sanitation systems and services sectors, most likely from industrial pollutants [56]. By modifying and switching up the geochemical conditions in the aquifer material, trace metals can dissolve in recharge water. The system's new redox conditions are brought about by the recharge water's abundance in nutrients and organic matter, which is driven by the microbial community [50, 55].

Several countries have created criteria on the recharging of treated wastewater such as in the United States and Australia. These guidelines highlight the health and environmental risks resulting from the occurrence of pathogenic microorganisms, suspended solids, and dissolved organic carbon in the water. There aren't many guidelines for trace element concentrations in water, except for those for five trace metals: lead, arsenic, nickel, cadmium, and mercury [47]. In addition, arsenic and nitrate are the main pollutants with severe hazards to public health. High levels of arsenic in groundwater have been realized as a great public health concern in many countries, which often resulted from natural origins instead of anthropogenic activities [57]. The existence of higher amounts of arsenic in drinking water is fundamental in Asian countries such as Bangladesh, India, and China, and recently, the same issue has been documented in European regions, United States, and Australia. For instance, in Bangladesh, groundwater contamination by arsenic is considered the greatest poisonous to the population in history [58]. It was reported that in the Bengal Delta, which includes India and Bangladesh, a population of 36 million might have polluted water to drink with arsenic. Ng et al. [56] discussed the impacts of long-term exposure to arsenic in drinking water on human health. Leading to an increase in the level of dissolved arsenic in groundwater could be raised due to the geochemical processes in the existence of oxygen dissolve arsenopyrite [FeAsS]. Oxidation can be a main factor for mobilization of the arsenic which is already present in aquifer rocks and can be increased due to refilling with oxygenated waters or by lowering the groundwater level through heavy pumping. Additionally, chemical interactions involving oxygen, nitrate, and iron can increase arsenic mobilization in groundwater [57, 59].

Moreover, groundwater contamination with nitrates is a diffuse and universal problem in both developed and developing countries. This problem can result from the excessive usage of fertilizers or animal waste and inappropriate municipal and animal waste disposal. The elevated level of nitrate in municipal groundwater (10–50 mg/l) is considered a hazard to public health. Foster and Chilton [60] showed that Nitrate contamination of groundwater as a result of agrochemicals has become a severe issue in China and India.

(B) **Emerging Pollutants**

Emerging pollutants and pathogens, such as antibiotic-resistant bacteria and viruses, currently threaten water quality and societal well-being. The recharge water that is mostly pharmaceutical-contaminated is treated sewage. In fact, the pharmaceutical compounds are released into water bodies from a variety of sources. For instance, the primary cause of wastewater contamination is patients' post-ingestive excretion of medicinal goods [55]. Therefore, wastewater from hospitals contains a higher concentration of pharmaceutical products such as antibiotics, anaesthesia products, disinfectants, and diagnostic products. From an eco-toxicological standpoint, wastewater may contain active pharmaceutical compounds that have been found to be harmful to the environment, as well as their byproducts, if they are active. Depending on the required quality and the characteristics of the treatment methods used in the treatment plant, different pharmaceutical compounds and/or their byproducts are removed from wastewater in different ways [59]. Moreover, the disposal of pharmaceutical products does not mean their complete eradication, as they can be degraded into other active products [61].

To solve this problem, pharmaceutical products' mobilization can be declined with infiltration basins and indirect ways of injection, similar to that of trace metals as the existence of an unsaturated zone promotes pollutant trapping. Through (1) adsorption and (2) biodegradation, microbiological and geochemical mechanisms can reduce the amount of hazardous or harmless pathogens in the unsaturated region.

The dispersion of organic pollutants can be significantly decreased by adsorption. Many solid states in the unsaturated zone have the ability to adsorb pharmaceuticals, such as humic substances mineralogical clays, and oxyhydroxides. In order for pharmaceutical substances to be adsorbed, electrical or chemical connections must be formed between their active units and those of the solid state in the unsaturated zone. The release of chemical compounds to the aqueous phase may or may not occur as a result of the adsorption of these pharmaceutical products [46].

(C) **Groundwater Contamination by Microorganisms**

The contamination of groundwater with microorganisms can occur due to appropriate protection of aquifers against releasing sewage effluent into groundwater. Groundwater contamination can occur in various ways, such as from rural on-site sewage services, septic tank and sewers leaks, and municipal dumps close to recharging locations. Water passing through the unsaturated zone can effectively decrease the concentration of many hazardous bacteria; however, the extent of pathogenic removal differs based on soil type, pollution degree, and kind of pollution. In the unsaturated zone, the natural reduction is often very effective, especially in the topsoil layers where microbial activity is higher. Human beings and animals waste include a variety of microbiological and viruses that pollute groundwater and create health issues. In 2012, the number of deaths from diarrhea was estimated to be more than 500,000, which were caused by microbial-contaminated drinking water [57]. According to a recent holistic overview by Ngure et al. [62] presents an evaluation of the microbiological pollution of water for human consumption on a worldwide scale. Since microbiological contamination of water may happen in freshwater sources as well as transmission lines, so groundwater cannot be attributed for all cases of contamination. Nonetheless, considering that over 50% of people globally obtain their drinking water from groundwater that may be polluted at some point, a sizable portion of outbreaks of waterborne diseases might be linked to groundwater [46].

6.3 Socio-economic Impacts

Public-private collaborations are essential to the accomplishment of neighborhood initiatives like MARs. MAR initiatives ought to be inclusive and collaborative from the outset to the finish. As part of their corporate social responsibility, private companies should offer financial and technical help to governments and stakeholders. Farmers are the largest water consumers in the MENA region, so they should receive special attention. Over the past few decades, a number of MAR projects in the MENA region have been impeded by farmers' denials. According to Niazi et al. [63], even though it wasn't financially feasible, the most socially acceptable scenario for MAR in the Sirik region of Iran was the most sustainable one.

Rather than closing down farmers' wells, it would let water to naturally flow into boreholes on agricultural property and allow farmers to upgrade their wells at no expense. Numerous researches revealed that obtaining societal acceptance was impossible in a number of Middle East and North Africa (MENA) MAR initiatives. For example, the economic analysis of an aquifer in Muscat, Oman, showed that injecting treated sewage effluent was attractive; however, the local users rejected the idea of combining treated sewage effluent with the available water resources. Their primary limitation pertained to injecting treated sewage effluent. Customers in Tunisia also objected to purchasing oranges and olives from orchards watered with poorly treated wastewater [64]. Even after local monitoring, farmers in Tunisia, the West Bank (Palestine), and Qatar believed that injecting treated sewage effluent for agricultural purposes was dangerous [65].

However, there is a dearth of MAR economic analysis in MENA nations. The degree of sophistication of the appropriate MAR technology must be determined for economic analysis, and this depends heavily on the hydrogeological conditions as well as the values of land and water [66]. Many MENA countries do not have easy access to this information. For the sake of future research and MAR implementation guidelines, it is crucial to record capital and operating expenditures. Furthermore, the erratic nature of the regional market deters farmers in many MENA nations from participating in MAR projects, as farm revenues are highly unstable. Cost-benefit evaluations are crucial because MAR investments carry a high risk and do not yield immediate returns.

Examples from other dry regions show that MAR costs can differ significantly depending on the project size, location, hydrogeological conditions, chosen technology, and other factors. The economic statistics for 21 MAR locations in the arid regions of Australia and the United States were examined by Ross and Hasnain [67]. They demonstrated the spreading method's cost-effectiveness (US\$0.156/m^3) in comparison to aquifer storage, transfer, and recovery (US\$2.67/m^3). Aquifer storage and recovery/aquifer storage transfer, and recovery operations contributed 52 Mm^3 of water annually in certain regions of Australia, according to research done in 2008 by Peterdillon et al. [68]. It was also necessary to calculate the average leveled cost, which is the annual income required repaying all associated capital, operating, and maintenance costs throughout the course of the project divided by the supply volume.

According to the study's findings, big initiatives are more cost-effective than small ones. For minor projects, the anticipated average leveled costs were US$3/m^3, while for large projects; they were US$1.21/m^3. Both small and large projects have yearly recovery capacity of 15,000–75,000 and 75,000–2,000,000 m^3, respectively [68].

6.4 Health Risks

Ajjur and Baalousha [6] demonstrated that eight MENA countries (Morocco, Tunisia, Egypt, Palestine, Oman, Kewait and Qatar) utilized treated sewage effluent as a source of water in MAR projects, and Thirteen MENA nations used rainfall for the same purpose including Algeria, Tunisia, Egypt, Palestine, Israel, Jordon, Syria, Lebanon, Iraq, Iran, Bahrein, Qatar, and UAE. If the water replenished was with poor quality, the treated sewage effluent and the collected rain might pose health risks. The poor quality treated sewage effluent that infiltrated the spreading project in the Korba aquifer (Tunisia) degraded the recharged water. This contamination was caused by leaks from nearby water tanks and debris from improperly maintained rooftops [69]. Agriculture and human activities near MAR sites have polluted the recharged water in certain MAR installations in Jordan [70].

The health of humans is at danger from a variety of contaminants found in wastewater, including heavy metals, organic chemicals, notably endocrine disrupting ones, and pathogenic bacteria. Pathogenic bacteria are widely regarded as the most significant threat to human health among these factors. Wastewater is home to a wide range of pathogenic microorganisms, including bacteria (such as enteropathogenic Escherichia coli, Shigella spp., and Salmonella spp.), viruses (such as rotavirus, adenovirus, poliovirus, and hepatitis A virus), protozoans (such as Entamoeba histolytica, Giardia intestinalis [formerly G. lamblia], and Cryptosporidium parvum), and parasitic worms (e.g. Ascaris lumbricoides, Necator americanus, and Trichuris trichiura). Typhoid (Salmonella spp.), dysentery (Shigella spp. and E. hystolytica), gastroenteritis (enteropathogenic E. coli), diarrhoea, vomiting or malabsorption (adenovirus, rotavirus, C. parvum, G. lambila, and T. trichiura), cholera (V. cholera), ascariasis (A. lumbricoides), and anaemia (N. americanus) are just a few of the symptoms/diseases linked to such infections [71]. The health of the source population determines the pathogen concentrations in wastewaters. Additionally, certain populations are more or less susceptible to infection from diseases that trigger an immunological response. Because they are frequently exposed to enteric viruses from an early age, persons in underdeveloped and developing nations, for instance, typically show stronger immunity to these infections than people in industrialised countries [72].

"It is estimated that water contamination is responsible for 190 million sick cases and 60,000 deaths in China. According to the World Bank, such illnesses cost the government $23 billion annually or 1% of China's gross domestic product (GDP). Additionally, the effect on China's ecosystems and food supply is not taken into account." In 2013, business owners in Shandong province were accused of contaminating shallow groundwater due to wastewater disposal through injection wells, which ignited a firestorm on the Internet [73].

7 Conclusions

Aquifer recharge by wastewater has various risks on the environment, human health, society and economy if not managed by the convenient methods. Aquifers are a part of bigger social-ecological systems, and these systems' constituents produce a variety of feedbacks that affect the aquifer's condition. Groundwater quality degradation and accidental groundwater contamination induce socio-political disorders from the public. One of the ways for managing water resources both quantitatively and qualitatively and adapt to climate change is effective MAR management. Therefore, to manage the MAR, MENA nations ought to take into account utilizing high quality water sources that meet the WHO quality criteria. In addition to manage the MAR systems appropriately and monitoring the quality of recharged water to avoid the negative health impacts.

We can summarize the advantages of using treated wastewater as aquifer recharge in MENA countries:

- The source of extra water for irrigation.
- The preservation of pure water for later advantageous applications.
- An inexpensive source of water.
- Low-cost method of wastewater disposal that also guards against hygienic issues and contamination.
- A steady and consistent supply of water.
- The efficient use of nitrogen and phosphorus, two plant nutrients found in wastewater.
- Offers further wastewater treatment prior to groundwater replenishment.

8 Recommendations

Water levels and water quality should be observed in the region of MAR projects. Developing societies can utilize low-cost tracers (such as dyes and chloride) to assess the status of recharged water, whereas developed countries can use radon and stable isotopes of water. Periodic monitoring of MAR projects aids in detecting evapotranspiration and lateral flow water losses, as well as water degradation caused by diffuse groundwater.

References

1. Rateb A, Scanlon BR, Fakhreddine S (2022) How severe is water stress in the MENA Region? Insights from GRACE and GRACE-FO satellites and global hydrological modeling. In: Applications of space techniques on the natural hazards in the MENA region. Springer International Publishing, Cham, pp 51–65. https://doi.org/10.1007/978-3-030-88874-9_4
2. Mualla W (2018) Water demand management is a must in MENA countries… but is it enough. J Geol Resour Eng 6:59–64. https://doi.org/10.17265/2328-2193/2018.02.002
3. Sowers J, Vengosh A, Weinthal E (2011) Climate change, water resources, and the politics of adaptation in the Middle East and North Africa. Climatic Change 104:599–627. https://doi.org/10.1007/s10584-010-9835-4
4. Van Steenbergen F, El Haouari N (2011) The blind spot in water governance: conjunctive groundwater use in the MENA countries. Water policy and law in the Mediterranean: an evolving nexus. UNESCO, Paris
5. Haddaoui I, Mateo-Sagasta J (2021) A review on occurrence of emerging pollutants in waters of the MENA region. Environ Sci Pollut Res 28(48):68090–68110. https://doi.org/10.1007/s11356-021-16558-8
6. Ajjur SB, Baalousha HM (2021) A review on implementing managed aquifer recharge in the Middle East and North Africa region: methods, progress and challenges. Water Int 46(4):578–604. https://doi.org/10.1080/02508060.2021.1889192
7. Sherif M, Sefelnasr A, Al Rashed M, Alshamsi D, Zaidi FK, Alghafli K et al (2023) A review of managed aquifer recharge potential in the Middle East and North Africa Region with examples from the Kingdom of Saudi Arabia and the United Arab Emirates. Water 15:742. https://doi.org/10.3390/w15040742
8. Kellis M, Kalavrouziotis I, Gikas P (2013) Review of wastewater reuse in the Mediterranean countries, focusing on regulations and policies for municipal and industrial applications. Glob NEST J 15(3):333–350
9. Rahman MA, Rusteberg B, Uddin MS, Saada MA, Rabi A, Sauter M (2014) Impact assessment and multicriteria decision analysis of alternative managed aquifer recharge strategies based on treated wastewater in Northern Gaza. Water 6(12):3807–3827. https://doi.org/10.3390/w6123807
10. Powley HR, Durr HH, Lima AT, Krom MD, Van Cappellen P (2016) Direct discharges of domestic wastewater are a major source of phosphorus and nitrogen to the Mediterranean Sea. Environ Sci Technol 50(16):8722–8730. https://doi.org/10.1021/acs.est.6b01742
11. Zyadin A (2013) Water shortage in MENA Region: an interdisciplinary overview and a suite of practical solutions. J Water Resour Prot 5(04):49. https://doi.org/10.4236/jwarp.2013.54A008
12. Asaad S, Suleiman A (2023) Socio-demographic factors and treated wastewater reuse in the MENA region: insights and implications. Desalination 565:116830. https://doi.org/10.1016/j.desal.2023.116830
13. El-Rawy M, Makhloof AA, Hashem MD, Eltarabily MG (2021) Groundwater management of quaternary aquifer of the Nile Valley under different recharge and discharge scenarios: a case study Assiut governorate, Egypt. Ain Shams Eng J 12:2563–74. https://doi.org/10.1016/j.asej.2021.02.023
14. Hassan WH, Nile BK, Mahdi K, Wesseling J, Ritsema C (2021) A feasibility assessment of potential artificial recharge for increasing agricultural areas in the kerbala desert in iraq using numerical groundwater modeling. Water 13:3167. https://doi.org/10.3390/w13223167
15. El-Rawy M, Al-Maktoumi A, Zekri S, Abdalla O, Al-Abri R (2019) Hydrological and economic feasibility of mitigating a stressed coastal aquifer using managed aquifer recharge: a case study of Jamma aquifer, Oman. J Arid Land 11:148–59. https://doi.org/10.1007/s40333-019-0093-7
16. Al-Maktoumi A, Zekri S, El-Rawy M, Abdalla O, Al-Abri R, Triki C, Bazargan L, Mohammad R (2020) Aquifer storage and recovery, and managed aquifer recharge of reclaimed water for management of coastal aquifers. Desalin Water Treat 176:67–77. https://doi.org/10.5004/dwt.2020.25499

17. Al-Maktoumi A, El-Rawy M, Zekri S (2016) Management options for a multipurpose coastal aquifer in Oman. Arab J Geosci 9:1–14. https://doi.org/10.1007/s12517-016-2661-x
18. Bekele E, Page D, Vanderzalm J, Kaksonen A, Gonzalez D (2018) Water recycling via aquifers for sustainable urban water quality management: current status, challenges and opportunities. Water 10:457. https://doi.org/10.3390/w10040457
19. Al Maktoumi A, Kassimov A, El-Rawy M (2022) Managed aquifer recharge as a reliable and sustainable water supply in Duqm area, Sultanate of Oman. 138-143. Abstract from The International Conference "Water Resources Management and Sustainability: Solutions for Arid Regions", Dubai, United Arab Emirates
20. Djellali M, Guefaïfia O, Fehdi C, Djellali A, Hamad A (2023) Research paper assessing the impact of artificial recharge on groundwater in an over-exploited aquifer: a case study in the Cheria Basin, North-East of Algeria. J Groundw Sci Eng 11:277
21. Abdelrady A, Sharma S, Sefelnasr A, El-Rawy M, Kennedy M (2020) Analysis of the performance of bank filtration for water supply in arid climates: case study in Egypt. Water 12(6):1816. https://doi.org/10.3390/w12061816
22. Yuan C, Wei Y, Xu X, Cao X (2023) Transport and transformation of arsenic in coastal aquifer at the scenario of seawater intrusion followed by managed aquifer recharge. Water Res 229:119440. https://doi.org/10.1016/j.watres.2022.119440
23. Bachtouli S, Comte J-C (2019) Regional-scale analysis of the effect of managed aquifer recharge on saltwater intrusion in irrigated coastal aquifers: long-term groundwater observations and model simulations in NE Tunisia. J Coast Res 35:91–109. https://www.jstor.org/stable/26568597
24. Jeuland M (2015) Challenges to wastewater reuse in the Middle East and North Africa. Middle East Dev J 7:1–25. https://doi.org/10.1080/17938120.2015.1019293
25. Kabour A, Hani A, Chebbah L, Maazouzi A (2015) Wastewater discharge impact on groundwater quality of Bchar City, Southwestern Algeria: approach by microbiological analysis and health risk index evaluation. J Eng Technol Res 7:44–52
26. Alnashiri HM (2021) Assessment of physicochemical parameters and heavy metal concentration in the effuents of sewage treatment plants in Jazan Region, Saudi Arabia. J King Saud Univ Sci 33:101600. https://doi.org/10.1016/j.jksus.2021.101600
27. Hassen W, Mehri I, Beltifa A, Giorgia Potortì A, Khellaf N, Amer R et al (2022) Chemical and microbiological assessment of wastewater discharged along the Mediterranean Sea. Sustainability 14(5):2746. https://doi.org/10.3390/su14052746
28. Choukr-Allah R (2010) Innovative wastewater treatments and reuse technologies adapted to southern Mediterranean countries. In: Waste water treatment and reuse in the Mediterranean region, pp 29–41. https://doi.org/10.1007/978-3-642-18281-5
29. Mostafa H, El-Gamal F, Shalby A (2005) Reuse of low quality water in Egypt. In: Non-conventional water use: WASAMED project Bari: CIHEAM/EU DG research (Options Méditerranéennes: Série B Etudes et Recherches), pp 93–103
30. Department of Environmental Protection PU (2012) Reuse of treated wastewater guidance manual. In: Pennsylvania. https://bfenvironmental.com/documents/bf-environmental-septic-water-reuse-manual.pdf
31. Abdelmoula S, Sorour MT, SaA Aly (2021) Cost analysis and health risk assessment of wastewater reuse from secondary and tertiary wastewater treatment plants. Sustainability 13:13125. https://doi.org/10.3390/su132313125
32. Mara D (2008) Waste stabilization ponds: a highly appropriate wastewater treatment technology for Mediterranean countries. In: Efficient management of wastewater. Springer, pp 113–123. https://doi.org/10.1007/978-3-540-74492-4_10
33. Al-Sa'ed R, Sayadi S, Ghata A et al (2009) Advancing membrane technologies for wastewater treatment and reclamation in selected Arab MENA countries. Desalin Water Treat 4:287–93. https://doi.org/10.5004/dwt.2009.496
34. Tropp H, Jagerskog A (2006) Water scarcity challenges in the Middle East and North Africa (MENA). Human Development Paper. UNDP. http://hdr.undp.org/hdr2006/pdfs/background-docs/Thematic_Papers/SIWI.pdf

35. Abu-Madi MOR (2004) Incentive systems for wastewater treatment and reuse in irrigated agriculture in the MENA region, evidence from Jordan and Tunisia. CRC Press
36. Kalavrouziotis IK, Arslan-Alaton I (2008) Reuse of urban wastewater and sewage sludge in the Mediterranean countries: case studies from Greece and Turkey. Fresenius Environ Bull 17:625
37. El-Rawy M, Zlotnik VA, Al-Raggad M, Al-Maktoumi A, Kacimov A, Abdalla O (2016) Conjunctive use of groundwater and surface water resources with aquifer recharge by treated wastewater: evaluation of management scenarios in the Zarqa River Basin, Jordan. Environ Earth Sci 75:1146. https://doi.org/10.1007/s12665-016-5946-1
38. Amin HM, Gad A, El-Rawy M, Abdelghany UA, Sadeek RA (2022) Improvement of partially treated wastewater quality by soil aquifer treatment in upper Egypt. J Eng Sci Technol 17:0689–0712
39. Missimer TM, Maliva RG, Ghaffour N, Leiknes T, Amy GL (2014) Managed aquifer recharge (MAR) economics for wastewater reuse in low population Wadi Communities, Kingdom of Saudi Arabia. Water 6:2322–38. https://doi.org/10.3390/w6082322
40. Bouri S, Dhia HB (2010) A thirty-year artificial recharge experiment in a coastal aquifer in an arid zone: the Teboulba aquifer system (Tunisian Sahel). Comptes Rendus Geosci 342:60–74. https://doi.org/10.1016/j.crte.2009.10.008
41. Qadir M, Bahri A, Sato T, Al-Karadsheh E (2010) Wastewater production, treatment, and irrigation in Middle East and North Africa. Irrig Drain Syst 24:37–51. https://doi.org/10.1007/s10795-009-9081-y
42. Usali N, Ismail MH (2010) Use of remote sensing and GIS in monitoring water quality. J Sustain Dev 3:228. https://doi.org/10.1007/978-3-031-40447-4_9
43. Custodio E (2013) Trends in groundwater pollution: loss of groundwater quality & related services. Groundwater Governance: a Global Framework for Action, GEF-World Bank-UNESCO http://www.groundwatergovernance
44. El Arabi NE (1999) Problems of groundwater quality related to the urban environment in Greater Cairo. IAHS Publication, pp 29–38
45. Amer R, Ripperdan R, Wang T, Encarnación J (2012) Groundwater quality and management in arid and semi-arid regions: case study, Central Eastern Desert of Egypt. J African Earth Sci 69:13–25. https://doi.org/10.1016/j.jafrearsci.2012.04.002
46. Jakeman AJ, Barreteau O, Hunt RJ, Rinaudo J-D, Ross A (2016) Integrated groundwater management. Springer Nature. https://doi.org/10.1007/978-3-319-23576-9
47. Bouwer H (1998) Groundwater and recharge principles. In: Proceedings of the third international symposium on artificial recharge of groundwater (TISAR 98)
48. El-Fakharany Z (2013) Environmental Impact assessment of artificial recharge of treated wastewater on groundwater aquifer system case study Abu-Rawash, Egypt. J Am Sci 9:309–15
49. El-Arabi N, Rashed M (1996) Vermeulen A. Environmental impacts of sewage water irrigation on groundwater. In: Proceedings of the 16th ICID congress on irrigation and drainage "Managing Environmental Changes Due to Irrigation and Drainage", pp 102–114
50. Pettenati M, Croiset N, Picot-Colbeaux G et al (2012) Optimisation of wastewater treatments through combined geomaterials and natural soil filter: modelling tools. J Water Reuse Desalin 2:185–93. https://doi.org/10.2166/wrd.2012.023
51. Hamilton AJ, Boland A-M, Stevens D et al (2005) Position of the Australian horticultural industry with respect to the use of reclaimed water. Agric Water Manag 71:181–209. https://doi.org/10.1016/j.agwat.2004.11.001
52. Hamilton AJ, Versace VL, Stagnitti F et al (2006) Balancing environmental impacts and benefits of wastewater reuse. WSEAS Trans Environ Dev 2:117–29
53. Anderson EI (2003) An analytical solution representing groundwater–surface water interaction. Water Resour Res 39. https://doi.org/10.1029/2002WR001536
54. Haeber R, Waller D (1987) Water quality of rainwater collection systems in the Eastern Caribbean. In: Proceedings of the third international rainwater catchment systems conference proceedings. http://www.eng.warwick.ac.uk/ircsa
55. Casanova J, Devau N, Pettenati M (2016) Integrated groundwater management, managed aquifer recharge: an overview of issues and options. ISBN:978-3-319-23575-2

56. Cauchi H, Nakache S, Zagury B, Carre C, Denis D, Larbaigt D, Martigne S (1996) Dossier: la réutilisation des eaux usées après épuration. Techniques. Sciences et méthodes 2:81–118
57. Fienen MN, Arshad M (2016) The international scale of the groundwater issue. In: Jakeman AJ, Barreteau O, Hunt RJ, Rinaudo JD, Ross A (eds) Integrated groundwater management. Springer, Cham. https://doi.org/10.1007/978-3-319-23576-9_2
58. Smith AH, Lingas EO, Rahman M (2000) Contamination of drinking-water by arsenic in Bangladesh: a public health emergency. Bull World Health Organiz 78:1093-103. https://www.ncbi.nlm.nih.gov/pmc/articles/PMC2560840/pdf/11019458.pdf
59. Jim TY, Bouwer EJ, Coelhan M (2006) Occurrence and biodegradability studies of selected pharmaceuticals and personal care products in sewage effluent. Agric Water Manag 86:72–80. https://doi.org/10.1016/j.agwat.2006.06.015
60. Foster S, Chilton P (2003) Groundwater: the processes and global significance of aquifer degradation. Philos Trans R Soc London Ser B: Biol Sci 358:1957–72. https://doi.org/10.1098/rstb.2003.1380
61. Zwiener C, Seeger S, Glauner T, Frimmel F (2002) Metabolites from the biodegradation of pharmaceutical residues of ibuprofen in biofilm reactors and batch experiments. Anal Bioanal Chem 372:569–75. https://doi.org/10.1007/s00216-001-1210-x
62. Ngure FM, Reid BM, Humphrey JH, Mbuya MN, Pelto G, Stoltzfus RJ (2014) Water, sanitation, and hygiene (WASH), environmental enteropathy, nutrition, and early child development: making the links. Ann N Y Acad Sci 1308:118–28. https://doi.org/10.1111/nyas.12330
63. Niazi A, Prasher SO, Adamowski J, Gleeson T (2014) A system dynamics model to conserve arid region water resources through aquifer storage and recovery in conjunction with a dam. Water 6:2300–21. https://doi.org/10.3390/w6082300
64. Hussain MI, Muscolo A, Farooq M, Ahmad W (2019) Sustainable use and management of non-conventional water resources for rehabilitation of marginal lands in arid and semiarid environments. Agric Water Manag 221:462–76. https://doi.org/10.1016/j.agwat.2019.04.014
65. Dare A, Mohtar RH (2018) Farmer perceptions regarding irrigation with treated wastewater in the West Bank, Tunisia, and Qatar. Water Int 43:460–71. https://doi.org/10.1080/02508060.2018.1453012
66. Dillon P, Arshad M (2016) Managed aquifer recharge in integrated water resource management. In: Integrated groundwater management. Springer, Cham, pp 435–452. https://doi.org/10.1007/978-3-319-23576-9_17
67. Ross A, Hasnain S (2018) Factors affecting the cost of managed aquifer recharge (MAR) schemes. Sustain Water Resour Manag 4:179–90. https://doi.org/10.3390/w14203257
68. Peterdillon IG, Contreras S, Pavelic P (2009) Managing aquifer recharge and discharge to sustain irrigation livelihoods under water scarcity and climate change. In: Improving integrated surface and groundwater resources management in a vulnerable and changing world, p 1
69. Abdul-Hamid M, Mahmoud N (2008) Rainwater harvesting for domestic uses in two Palestinian Rural Areas with emphasis on quality and quantity. Birzeit University. http://hdl.handle.net/20.500.11889/1715
70. Xanke J, Liesch T, Goeppert N, Klinger J, Gassen N, Goldscheider N (2017) Contamination risk and drinking water protection for a large-scale managed aquifer recharge site in a semi-arid karst region, Jordan. Hydrogeol J 25:1795–809. https://doi.org/10.1007/s10040-017-1586-0
71. Yates DN, Strzepek KM (1998) An assessment of integrated climate change impacts on the agricultural economy of Egypt. Climatic Change 38:261–87
72. Shuval HI (1991) The development of health guidelines for wastewater reclamation. Water Sci Technol 24:149–55. https://doi.org/10.2166/wst.1991.0195
73. Liu, J., Zheng, C (2016) Towards integrated groundwater management in China. In: Jakeman AJ, Barreteau O, Hunt RJ, Rinaudo JD, Ross A (eds) Integrated groundwater management. Springer, Cham. https://link.springer.com/chapter/10.1007/978-3-319-23576-9_18https://link.springer.com/chapter/10.1007/978-3-319-23576-9_18

The Impact of Aquifer Recharge on Groundwater Quality

Ahmed Khaled Abdella Ahmed and Mustafa El-Rawy

Abstract High-quality water is rare in arid and semi-arid regions like those found in the MENA countries. The hydrological cycle involves groundwater recharge, wherein water moves from the water table to the saturated zone through percolation or drainage, entering the aquifer and flowing through the vadose zone. Both natural groundwater recharge via the hydrological cycle and artificial recharge are considered mechanisms for retaining water within the subsoil. Storing water in surface reservoirs presents drawbacks, such as significant evaporation losses, the need for extensive land, sediment accumulation, structural failure risks, and susceptibility to pollution. Consequently, there is a growing interest in exploring diverse Managed Aquifer Recharge (MAR) technologies to function as extensive reservoirs, offering effective alternatives in various hydrogeological contexts. This chapter aims to elucidate the various types of AR techniques and their influence on groundwater quality, highlighting their potential as sustainable and efficient approaches in water management across diverse domains. MAR using Treated Wastewater (TWW) can be considered safe, provided that specific precautionary measures are implemented to ensure the quality of the effluent, and the discharge of Treated Wastewater is carefully regulated. This safeguards freshwater resources for both current and future generations.

Keywords Groundwater quality · Aquifer recharge · Pollution · Bank infiltration · Rainfall harvesting · Soil aquifer treatment

---

A. K. A. Ahmed (✉)
Sanitary and Environmental Engineering, Civil Engineering Department, Faculty of Engineering, Sohag University, Sohag, Egypt
e-mail: dr.ahmed_khaled@eng.sohag.edu.eg

M. El-Rawy
Civil Engineering Department, Faculty of Engineering, Minia University, Minia 61111, Egypt

Civil Engineering Department, College of Engineering, Shaqra University, Dawadmi 11911, Saudi Arabia

M. El-Rawy
e-mail: mustafa.elrawy@mu.edu.eg

M. El-Rawy and A. Negm (eds.), *Managed Aquifer Recharge in MENA Countries*, Earth and Environmental Sciences Library,
https://doi.org/10.1007/978-3-031-58764-1_10

1 Introduction

Climate change, rapid urbanization, and population growth have all contributed to growing variations in water demand and supply [1–3]. In arid and semi-arid areas, where it may be the only source of freshwater is groundwater, which is essential for home and agricultural use, especially during times of high-water demand [4–6]. Coastal eco-hydrological environments and biogeochemical cycles are thought to be maintained by groundwater, which is now recognized as a key link in the hydrological cycle [7]. The water table is the term used to describe the top layer of saturated region underground. The ocean's level, which is immediately below the beach's surface, and the water table level at the beach are always the same [8].

Surface reservoirs are frequently used to store water; however, there are a number of drawbacks to this practice, including high evaporation losses, a need for a lot of land, the buildup of sediment, the likelihood of structural failure, and a high vulnerability to pollution [9]. During times of low demand or abundant availability, it is possible to store extra water underground as an alternative to surface storage. An effective strategy for managing water resources sustainably is aquifer recharge, which is the deliberate replenishment of an aquifer for future use or environmental advantages [10]. Aquifers are enormous reservoirs of water, and they are the saturated region below the water table [11]. Supporting increased water demand, resolving the temporal gap between water supply and demand, enhancing water quality, and protecting water from evaporation losses, algal blooms, and air pollution fallout are all important during subterranean storage are all substantial advantages of aquifer recharge over alternative methods of managing water supplies [12].

River water, stormwater, precipitation, desalinated saltwater, treated wastewater effluent [13–15], and even groundwater from other aquifers are some of the several types of water that can be stored in an appropriate aquifer. As a result, there is an increased understanding of its ability to manage water resources, which spurs the expansion of aquifer recharge (AR) globally. According to reports, AR implementation has reportedly advanced in recent years at a pace of 5% per year, reaching an estimated ten thousand km^3 every year at this point [16].

The past decades have seen unprecedented groundwater extraction and overdraft as well as development of new technologies for water treatment that together drive the advance in intentional groundwater replenishment known as aquifer recharge. Numerous researches on the topic of AR have been conducted with a variety of focuses, including many scientific journal publications, management reports, and technical assessments. As a result, various AR technologies and processes for AR implementation have been established [17–21]. Many studies [22] concentrated on the deterioration of groundwater quality across different global regions, especially in densely populated areas. They indicated that the principal challenges impacting groundwater arise from issues related to drainage systems and a deficiency in monitoring and controlling sewage systems effectively [22].

In arid and semi-arid regions like those found in MENA countries, the availability of high-quality water is restricted, and water of marginal acceptability remains underutilized. Adopting Treated Wastewater (TWW) for reuse is a viable strategic option. In areas with suitable hydrogeological conditions, aquifers can temporarily store TWW through artificial recharge, allowing for subsequent retrieval [23, 24]. Subsurface storage and recovery techniques are often more cost-effective and environmentally sustainable compared to conventional surface storage methods [17, 25]. However, the chemical composition of water stored in an aquifer during the MAR is may influenced by chemical reactions driven by the aquifer's conditions (such as pH, redox state, minerals, organic matter, and microbial activity) on the quality of the groundwater [26]. However, MAR using TWW can be considered safe, provided that specific precautionary measures are implemented to ensure the quality of the effluent, and the discharge of TWW is carefully regulated [18].

A sizable amount of research has been conducted recently to help increase our knowledge of the natural or artificial processes involved in aquifer recharge, the application of different AR technologies, and how to manage these systems more effectively in a variety of hydrogeological settings and domains. Thus, the current chapter's goals are to describe the different types of AR and their effect on underground water quality. Finally, this chapter also reveals a potential future direction for aquifer recharge.

2 Illustrative History of Aquifer Recharge

The lengthy history of AR is explained in numerous sources of peer-reviewed literature, along with examples of its use [16–18, 23–25, 27–29]. In addition to being known as aquifer recharge, AR is also known as artificial recharge, enhanced recharge, and water banking [30, 31]. The phrase "artificial recharge" was first mentioned in 1900 [32], and defined it as "subterranean" water replenished by human activity was previously the most often used one. Some authors claimed that the term "artificial recharge" was deceptive because it suggested that a procedure that purified the subsoil used artificial means rather than natural ones. Afterwards, Tuinhof et al. [33] presented the basic idea of "management aquifer recharge," which refers to the management of aquifer recharge and subsurface storage. Aquifer recharge (AR) was initially used in invented [34]. AR progressively spreads in society because "controlled "aquifer recharge" suggests that risks are handled quantitatively, [35]. Based on four stages, the historical development of AR may be divided on the rate of development and the objects served as follows:

(a) The first phase lasted mostly between 221 BC and 1850 AD. AR was still in its early stages, with little development and little use. The AR prototypes often functioned as a means of increasing groundwater level for irrigation of farmland. Channels were built in China during the Warring States period (475 BC–221 BC) to make it easier for surface water to seep into the ground and improve the quality of the groundwater. The Karez typically consists of a shaft, an under drain, an open channel, and a water-logging dam. A Karez was created during

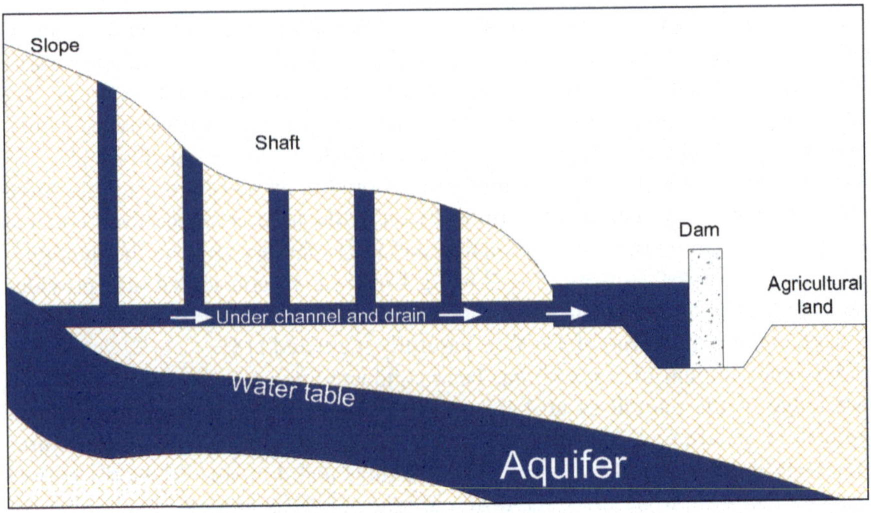

Fig. 1 Illustrative scheme of Karez. After [14]

the Qin and Han Dynasties, China (221 BC–220 AD) to capture groundwater for agricultural irrigation and domestic water use on low land [36], see Fig. 1. The Wari (a pre-Incan society) were said to have created "Amunas," one of the earliest theories of infiltration canals, for use in water supply and runoff. The highest portion of the Sierra Nevada in Spain was where the Arabs built careo (channels to filtrate water to underground), which allowed water to be infiltrated into the aquifer during peak flows, resulting in a more stable base flow for spring irrigation [37].

(b) The second phase occurred during the 19th century's rapid industrialization, when Europe's expanding population created enormous challenges to water supplies. The conventional inadequate sanitation and rising pollution from new enterprises have hurt water supplies that rely on surface water. AR and its use as a result became the remedy from 1850 and 1950. Theoretical, technological, and practical advancements in AR were made rather quickly. Darcy established the renowned Darcy's law in 1856, which laid the groundwork for the theoretical investigation of seepage calculation. The idea of steady-state flow in pumping wells was essentially developed in the 1930s after the Dupuit formula was proposed in 1863 [38]. Theis [39] first presented the theory of transient-state flow in pumping wells in 1935. The growth of theoretical research has been accompanied by the use of AR was successful in both Europe and America. In the lower Rhine, river bank filtration was implemented in 1870 and quickly spreaded to the UK, Finland and the Netherlands [40]. In California (USA), the first infiltration ponds for storm runoff were constructed in 1900 and then became popular in the 1930s. These techniques were effective in restoring the damaged surface water. Greater water supply and improved water quality to be available to industries demand [41].

(c) The third phase, after World War II, needs for water quantity and quality of water supply were advanced due to post-war reconstruction and urbanization-related activities, resulting in the third stage of AR. From 1950 to 1990, AR was increasingly widely used to meet the demands of the post-war era. It should be noted that AR continued to see tremendous expansion in both theoretical research and application after World War II. In the 1970s, researchers like Jacob, Boulton, and Hantush, among others, developed the theory of transient state flow in pumping wells. On the basis of the theory of groundwater flow in pumping wells, Huisman and Olsthoorn [42], for instance, derived the analytical solution for the recharging condition. AR projects abound in addition to these theoretical research initiatives [42]. The concepts behind Aquifer Storage and Recovery (ASR) emerged during this period. In order to prevent saline intrusion and help meet water demand during the peak season, the first long-term ASR well field was put into place in 1968 in Wildwood, New Jersey, USA [43]. Large-scale AR projects were started in Europe's Netherlands in the 1950s and Finland in the 1960s with the intention of supplying water [35]. On the Burdekin Delta and the Queenslans, AR project operations were started in Australia in the middle of the 1960s [44]. In order to stop the difficulties with ground sinking in Shanghai, China, in the 1960s, artificial recharge using wells was deployed. The aforementioned account explains how the idea of AR evolved and expanded among nations through research and practical application [38].

(d) The fourth phase is thought to have begun in the 1990s when the idea of AR is regarded as having been used in many nations to address the issues of water scarcity brought on by population growth and climate change, in addition to safeguarding the ecosystem and environment. During this time, the concept of AR was increasingly applied to research and construction-related tasks. For instance, the number of ASR programs expanded in various nations, including the USA [43]. The AR idea was being adopted in a discernible trend. For instance, AR was no longer restricted to rich nations but instead extended internationally. However, there was a difference: while developing nations like China, India, and South Africa focused on planning and developing macro-strategies, developed nations in Europe, America, and other regions were more concerned with compiling technical specifications based on case studies that they implemented. The feasibility of MAR techniques has been frequently investigated in numerous regions in MENA countries [17, 18, 24, 45]. The advancement of science and technology, where computer technology is widely used in the assessment, design, and operation of AR and such developments in the science and technology of AR cut across all countries, is another factor contributing to the divide between developed and developing countries [10]. In conclusion, it can be said that the history of AR has demonstrated how the idea of AR originated and expanded throughout the world and how the advancement of science and technology has helped to reduce the gap between developed and poor nations.

3 Impact of Different Aquifer Recharge techniques on Groundwater

Several instances of Aquifer recharge (AR) types could be naturally or artificially created, and the United Nations Educational, Scientific, and Cultural Organization (UNESCO) summarized experiences of AR in numerous countries [38]. Several research studies provide updated overviews of AR approaches [46–48]. There is general literature on AR and material on AR technology is divided into the following four key groups based on the recharge and storage method: bank filtration and in-channel alterations, runoff harvesting, spreading techniques, and well or shaft and borehole recharging are among the options, see Fig. 2. The sections below give a brief description of each of these techniques and their impact on groundwater quality:

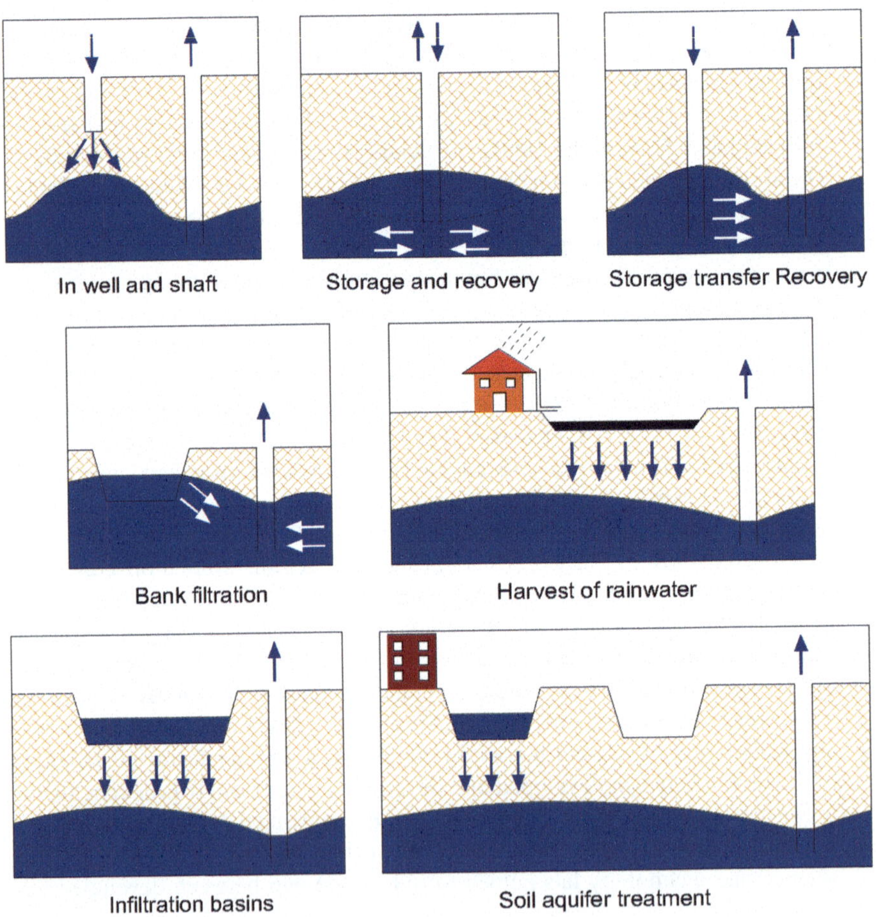

Fig. 2 Schematic diagram of different aquifer recharge types. Modified after [34]

3.1 Bank Infiltration (BI) and In-Channel Alterations

Dune filtration is an indirect recharging technique that is included in bank infiltration. In order to encourage infiltration from a surface water body and improve and make the quality of the water recovered more consistent, the bank filtration technique involves taking groundwater from a well or caisson close to or beneath a river or lake. Bank filtration and dunes filtration are comparable [49]. The rivers, streams, or canals are modified using in-channel alterations procedures in order to increase vertical recharge and store water. Percolation ponds, sand storage dams, underground dams, leaky dams, recharge releases, and other types of dams are used to provide in-channel changes [50].

Bank infiltration (BI) could enhance or impair the aquifer groundwater quality. The following benefits of bank filtration are achieved through its efficient natural attenuation processes: elimination of suspended solids, particles, biodegradable compounds, bacteria, viruses, and parasites; partial elimination of adsorbable compounds; and equilibration of temperature changes and concentrations of dissolved constituents in the bank filtrate. Increases in hardness, ammonium, dissolved iron, and manganese concentrations, as well as the creation of hydrogen sulfide and other foul-smelling Sulphur compounds as a result of altering redox conditions, are all undesirable effects of bank filtration on water quality [51–53]. The intricately interconnected geochemical, biological, and hydrologic elements that affect how well the bank infiltration process works are complex. The efficiency of bank infiltration is affected by several factors, including; climate change, soil and water characteristics, water level Fluctuations, bank area uses, aquifer nature, hydraulic conductivity, and travel distance [1]. Some of these factors are discussed as follows.

Related to climate change, temperature fluctuates significantly in some rivers; this can have an impact on how quickly rivers degrade via the BI process. The Van't Hoff hypothesis states that higher temperatures are related to faster biological and chemical reactions [54]. High water temperatures can accelerate degradation in the BI process. However, continuing the stimulation over time is not recommended because it can cause anoxic or anaerobic conditions, which in turn slow down the pace of disintegration [55].

Soil type has valuable effects on water quality during BI process. The kind of soil influences microbial clearance. In assessing the risk of water pollution, knowledge and understanding of subsurface media's effectiveness on microbial elimination are essential. Significant microbial clearance rates are found in volcanic soils, fine sand, pumice sand, and heavily worn aquifer rocks. However, low removal rates have been noted in aquifers of coarse gravel, karst limestone, structured stony soils, and fractured rocks [56]. For usage with the BI method, natural clay composites, including montmorillonite, bentonite, and other clay minerals work well. These minerals' benefits include a large specific surface area and an important ability to capture exchangeable cations [57, 58].

The infiltrated water's subsurface velocity is directly impacted by river/lake water level changes. The water level fluctuates because of snowmelt, floods, hurricanes, or other climate changes [59]. Infiltrated water quantity and quality may alter as a result of the BI process. According to Derx et al. [60], there is a close relationship between changes in river water level and the river aquifer mixing zone. They indicated that variations in river water levels are to blame for the dilution of solute concentrations. Thus, the movement of nutrients was significantly impacted. They suggested that the analysis of mixing zones take into account variations in river water levels. When compared to viruses in steady water levels, viruses in natural or manmade rivers/lakes with variable water levels are more likely to penetrate riverbanks across greater distances and in larger concentrations. The viral concentrations increased by 2–4 logs with an increase in water level between 1 and 5 m, but the travel time was reduced by 30% [61].

River biological diversity and ecological processes are significantly impacted by land use, such as agricultural growth and urbanisation within the nearby valley of streams and rivers. River ecosystem characteristics such as water quality, community structure, primary and secondary production, organic matter decomposition, ecosystem metabolism, and energy flows can all be impacted by changes in land use in river basins [62, 63].

Hydraulic-based or tracer-based methods can be used to calculate travel times in the infiltration zone between a river and the pumping well(s). Hydraulic methods are used to measure the hydraulic conductivity, hydraulic gradients, and porosity of the sediments that are situated between the natural industrial river, lake, or pond bed and the well(s). Then, using calculations based on Darcy's law or numerical groundwater flow modelling, flow velocities and journey periods are inferred indirectly [64]. More precise flow velocity and trip time measurements might be made using tracer-based techniques. The extremely desired travel time characteristics of the BI process have a big influence on subsurface filtration reductions. A trip lasting longer than a month is necessary to ensure that natural processes sufficiently purify the surface water in the ground [49].

3.2 *Rainfall Harvesting*

Rainwater is gathered, routed through percolation into a large pit, and then reused. This method is effective for enhancing rainwater's natural filtering before entering aquifers. This procedure helps restoring water metropolitan areas' hydrological cycles function normally [65, 66].

It is impossible to separate the process of using rainwater harvesters to help supply water for daily requirements from the quality that has been specified by national standards and local regulations so that it is fit for usage. The qualities of rainwater that are utilised to recharge aquifers are as follows: (a) Though believed to be of high quality, harvested rainwater nevertheless picks up toxins along the way. This occurs as a result of the rain's air-dissolving action; (b) The ensuing effect of contaminants emerges in the first several weeks of rain due to the buildup of sediment, roofing material, animal faeces, and other contaminants in the air and media; (c) Only the first week of rain has the highest concentration of impurities; when it rains the next week, these contaminants have been washed away by the rain in the preceding weeks, resulting in good quality production; (d) In general, the use of roofing media has a capacity for absorption that offers less pollution than other media. In contrast to surface infiltration soil, precipitation that falls on the media can, therefore, preserve its quality [67]. As a result of non-point source contamination, the rainfall that is collected typically has poor water quality [68].

Regarding floods and droughts, water needs to be assessed and managed [69, 70]. Drought and flooding have a secondary impact on BI process effectiveness. Flooding reduces blockage of the streambed via scouring, whereas drought conditions have the reverse effect. Flooding causes a shear force known as self-purification, whereas a drought encourages the sedimentation of suspended materials, blocking the riverbed. The congestion of the riverbed can be both beneficial and unpleasant. On the one hand, it encourages pollutant biodegradation. On the other hand, it lessens the subsurface filtration zone's hydraulic conductivity [71]. When the BI path is dry, anaerobic conditions are boosted, encouraged, and reinforced. Flooding reduces travel time, stimulates the usage of the fast pass, and causes invasiveness of harmful micropollutants [55].

3.3 Spreading Techniques

Spreading techniques for aquifer recharge, such as infiltration ponds and basins, soil aquifer treatment (SAT), controlled flooding, treated sewage effluent, and excessive irrigation, are the most popular and practical ways to use. The most common infiltration ponds are water ponds (basins) that are either dug into the ground or open land that is surrounded by a bank and store water that will be recharged (such as stormwater, rainwater, or dam) with the intention of infiltrating that water into aquifers through permeable soils. Another common spreading technique is the soil aquifer treatment (SAT) system, which uses infiltration ponds to penetrate sporadically the soil. When an unconfined aquifer needs to be replenished, and it is possible to do so by infiltration through porous material, the water spreading method is used [38].

Spreading techniques have some merits and demerits on aquifer water quality. The merits include (a) The high effectiveness in enhancing water quality during emergency and drought periods; (b) Old or dry wells may be utilised; (c) Aquifer storage and recovery is favored in terms of economics. However, due to the longer infiltration duration in the aquifer storage, transfer, and recovery approach, aquifer storage, transfer, and recovery surpass aquifer storage and recovery if the primary purpose is water quality control and (d) No evaporation losses. The demerits include: (a) Injection capacity and recovery efficiency are influenced by a variety of parameters, including layout modifications, aquifer recharge capacity, native water quality, and discharge rate, which should be researched before use; (b) It is limited in brachitic or saline aquifers due to mixing with ambient water production and lowering recovery volume; (c) Pretreatment can be necessary to prevent blockage; and (d) Losses result from lateral flow.

In the spreading technique, surface water quality and soil layers characteristics are the dominant factors for aquifer recharge groundwater quality [72, 73]. The quality of surface water that infiltrates through land stratums plays an important role in groundwater quality during aquifer recharge. To avoid groundwater pollution, source water must be treated before the recharge process, based on the needed effluent quality and aquifer parameters, this treatment's target is determined [74]. Soil characteristics also have a valuable influence on groundwater quality during aquifer recharge through spreading techniques. Karst aquifer heterogeneity results in substantial blockage and effect dispersion, which lowers the recovery rate [75, 76]. Karst settings' water levels, water quality, and soil moisture should be watched closely around MAR projects. To track the fate of restored water, developing nations can employ less expensive tracers (such dyes and chloride), but radon and stable water isotopes are wise options. Potential evapotranspiration should be considered in aquifer recharge through spreading techniques. The quantity of water that could be transferred to the atmosphere under optimal soil moisture and vegetation conditions is known as potential evapotranspiration. The high evapotranspiration rates cause a loss of water and an increase in salt content [77].

3.4 Well, Shaft and Borehole Recharge

These techniques, which are commonly used in many nations where deep aquifers are frequent, include open wells and shafts, aquifer storage and recovery (ASR), and aquifer storage, transport, and recovery (ASTR). In these places, injection techniques are usually employed. Shallow phreatic aquifers are recharged via open wells and shafts. Spreading techniques typically fail in places with low permeability surface layers. Wells frequently dry up in these locations as a result of declining water tables brought on by over-exploitation. Water is stored below ground in ASR systems using the same wells for both injection and recovery. ASR has been enhanced, and ASTR refers to the injection and recovery of water from different wells. ASTR's prolonged residence length makes it efficient at enhancing the quality of stored water In order to properly manage the system, the wellfield and reservoir must be protected, and regular observation is also necessary [78].

A human activity known as irrigation has the power to change the quantity and quality of groundwater supplies [79]. A range of consequences of irrigation on groundwater supplies have been documented in earlier investigations in many different parts of the world [80], greater aquifer recharge in places that use surface water for irrigation [81]; groundwater depletion in areas where it feeds irrigation demand groundwater depletion in areas where it feeds irrigation demand [82, 83], and a decline in the quality of groundwater, especially with treated wastewater irrigation [84, 85]. Seawater, brackish water, contaminated surface waters, and mineral deposits are some of the natural sources of groundwater contamination. When human activities disturb the natural environmental balance, such as when aquifers are depleted and seawater intrusion occurs, when mineral resources are exploited and when excessive irrigation causes acid mine drainage, these natural sources may become significant sources of contamination [86].

4 Conclusions

High-quality water is scarce in arid and semi-arid regions such as those found in MENA countries, and water of borderline acceptability is underutilized. The increasing fluctuations in water demand and supply caused by climate change, rapid urbanization, and population increase highlight the crucial need for successfully managing groundwater resources. Given its critical role in providing home and agricultural water needs, protecting groundwater quantity and quality is essential. Aquifer Recharge (AR), whether natural or artificial, emerges as an important strategy in this endeavor. The purpose of this chapter is to clarify the various types of AR techniques and their impact on groundwater quality. This chapter categorized AR techniques into four categories: well or shaft and borehole recharge, runoff harvesting, spreading techniques, and bank filtration and in-channel changes, with each providing a unique means of improving groundwater recharge. The effectiveness of water recharge is closely related to a variety of parameters, including soil properties, water quality, climatic change, land use, and percolation travel time. Understanding and optimizing these parameters are important for developing long-term water management plans in the face of changing environmental and demographic issues. In order to protect groundwater from contamination, it is crucial to implement adaptive policies that optimize the potential of aquifer recharge while preventing groundwater pollution. This safeguards freshwater resources for both current and future generations. The successful realization of this goal requires adherence to adaptive policies and regulations governing managed aquifer recharge, along with a strict commitment to water quality standards.

5 Recommendations

As a recommendation, the Aquifer reservoir should be protected, and regular monitoring is also essential. Furthermore, preparation of surface water may be required to avoid obstruction and clogging and to ensure aquifer water quality. To optimize the effects of aquifer recharge on groundwater quality, it is vital to build integrated monitoring systems that follow water quality parameters throughout time. To limit contamination, best management practices should be implemented, with a focus on sustainable recharge techniques and land use planning. Natural attenuation processes within aquifers require research to guide the selection of recharge sites that improve self-purification mechanisms. Public awareness campaigns should educate communities, and coordination with policymakers is essential in developing and enforcing aquifer recharge guidelines.

References

1. Ahmed AK, Marhaba TF (2017) Review on river bank filtration as an in situ water treatment process. Clean Technol Environ Policy 19:349–359. https://doi.org/10.1007/s10098-016-1266-0
2. Page D, Vanderzalm J, Miotliński K, Barry K, Dillon P, Lawrie K, Brodie RS (2014) Determining treatment requirements for turbid river water to avoid clogging of aquifer storage and recovery wells in siliceous alluvium. Water Res 66:99–110. https://doi.org/10.1016/j.watres.2014.08.018
3. El-Rawy M, Batelaan O, Al-Arifi N, Alotaibi A, Abdalla F, Gabr ME (2023) Climate change impacts on water resources in arid and semi-arid regions: a case study in Saudi Arabia. Water 15(3):606. https://doi.org/10.3390/w15030606
4. Abdella AK, Abdel-Aa MH (2021) Iron removal from ground water through expanded polystyrene filter. J Environ Treat Tech 9(3):657–666
5. Gleeson T, Wada Y, Bierkens MFP, van Beek LPH (2012) Water balance of global aquifers revealed by groundwater footprint. Nature 488(7410):197–200. https://doi.org/10.1038/nature11295
6. El-Rawy M, Abdalla F, El Alfy M, Abdalla F (2020) Water resources in Egypt. In: Hamimi Z, El-Barkooky A, Martínez Frías J, Fritz H, Abd El-Rahman Y (eds) The geology of Egypt. Regional geology reviews. Springer, Cham. https://doi.org/10.1007/978-3-030-15265-9_18
7. David M, Bailly-Comte V, Munaron D, Fiandrino A, Stieglitz TC (2019) Groundwater discharge to coastal streams—A significant pathway for nitrogen inputs to a hypertrophic Mediterranean coastal lagoon. Sci Total Environ 677:142–155. https://doi.org/10.1016/j.scitotenv.2019.04.233
8. Baird AJ, Low RGJ (2022) The water table: Its conceptual basis, its measurement and its usefulness as a hydrological variable. Hydrol Process 36(6):e14622
9. Minsley BJ, Ajo-Franklin J, Mukhopadhyay A, Morgan FD (2011) Hydrogeophysical methods for analyzing aquifer storage and recovery systems. Ground Water 49(2):250–269. https://doi.org/10.1111/j.1745-6584.2010.00676.x
10. Ringleb J, Sallwey J, Stefan CJW (2016) Assessment of managed aquifer recharge through modeling—A review. Water 8(12):579
11. Bizhanimanzar M, Leconte R, Nuth MJH, Sciences ES (2019) Modelling of shallow water table dynamics using conceptual and physically based integrated surface-water–groundwater hydrologic models. Nat Comput Sci 23(5):2245–2260
12. Hartog N, Stuyfzand PJ (2017) Water quality considerations on the rise as the use of managed aquifer recharge systems widens. Environ Eng 9(10):808
13. Abdella Ahmed AKA, Shalaby M, Negim O, Abdel-Wahed T (2022) Egyptian reuse standards of treated wastewater for irrigation. J Environ Manag 47(2):43–59
14. El-Rawy M, Abd-Ellah MK, Fathi H, Ahmed AK (2021) Forecasting effluent and performance of wastewater treatment plant using different machine learning techniques. J Water Process Eng 44:102380. https://doi.org/10.1016/j.jwpe.2021.102380
15. Ahmed AK, El-Rawy M, Ibraheem AM, Al-Arifi N, Abd-Ellah MK (2023) Forecasting of groundwater quality by using deep learning time series techniques in an Arid Region. Sustainability 15(8):6529
16. Dillon P, Stuyfzand P, Grischek T, Lluria M, Pyne RDG, Jain RC, Sapiano M (2019) Sixty years of global progress in managed aquifer recharge. Hydrogeol J 27(1):1–30. https://doi.org/10.1007/s10040-018-1841-z

17. El-Rawy M, Al-Maktoumi A, Zekri S, Abdalla O, Al-Abri R (2019) Hydrological and economic feasibility of mitigating a stressed coastal aquifer using managed aquifer recharge: a case study of Jamma aquifer, Oman. J Arid Land 11:148–159. https://doi.org/10.1007/s40333-019-0093-7
18. Al-Maktoumi A, Zekri S, El-Rawy M, Abdalla O, Al-Abri R, Triki CMR, Bazargan-Lari MR (2020) Aquifer storage and recovery, and managed aquifer recharge of reclaimed water for management of coastal aquifers. Desalin Water Treat 1(176):67–77. https://doi.org/10.5004/dwt.2020.25499
19. Maliva RG (2014) Economics of managed aquifer recharge. Ground Water 6(5):1257–1279
20. Mukherjee A, Scanlon BR, Aureli A, Langan S, Guo H, McKenzie AA (2020) Global groundwater: source, scarcity, sustainability, security, and solutions. Elsevier
21. Shah T (2010) Climate change and groundwater: India's opportunities for mitigation and adaptation. Environ Res Lett 5(3):035002
22. El-Rawy M, Fathi H, Abdelrady A, Negm AM (2022) Environmental impacts of treated wastewater contaminates on groundwater quality in the Nile River Valley, Egypt. In: Sustainability of groundwater in the Nile Valley, Egypt. Springer International Publishing, Cham, pp 91–120. https://doi.org/10.1007/978-3-031-12676-5_5
23. Al-Maktoumi A, El-Rawy M, Zekri S (2016) Management options for a multipurpose coastal aquifer in Oman. Arab J Geosci 9:636. https://doi.org/10.1007/s12517-016-2661-x
24. El-Rawy M, Zlotnik VA, Al-Raggad M, Al-Maktoumi A, Kacimov A, Abdalla O (2016) Conjunctive use of groundwater and surface water resources with aquifer recharge by treated wastewater: evaluation of management scenarios in the Zarqa River Basin, Jordan. Environ Earth Sci 75:1–21. https://doi.org/10.1007/s12665-016-5946-1
25. Abdelrady A, Sharma S, Sefelnasr A, El-Rawy M, Kennedy M (2020) Analysis of the performance of bank filtration for water supply in arid climates: case study in Egypt. Water 12(6):1816
26. Page D, Bekele E, Vanderzalm J, Sidhu J (2018) Managed aquifer recharge (MAR) in sustainable urban water management. Water 10(3):239. https://doi.org/10.3390/w10030239
27. Alam S, Borthakur A, Ravi S, Gebremichael M, Mohanty SK (2021) Managed aquifer recharge implementation criteria to achieve water sustainability. Sci Total Environ 768:144992
28. Fakhreddine S, Prommer H, Scanlon BR, Ying SC, Nicot JP (2021) Mobilization of arsenic and other naturally occurring contaminants during managed aquifer recharge: a critical review. Environ Sci Technol 55(4):2208–2223
29. Amin HM, Gad A, El-Rawy M, Abdelghany UA, Sadeek RA (2022) Improvement of partially treated wastewater quality by soil aquifer treatment in upper Egypt. J Eng Sci Technol 17(1):0689–0712
30. Alam S, Gebremichael M, Li R, Dozier J, Lettenmaier DP (2020) Can managed aquifer recharge mitigate the groundwater overdraft in California's central valley? Water Resour Res 56(8):e2020WR027244
31. Dillon P, Fernández Escalante E, Megdal SB, Massmann G (2020) Managed aquifer recharge for water resilience. Water 12(12):1846
32. Richter JG (1900) On artificial underground water. Рипол Классик
33. Tuinhof A, Olsthoorn T, Heederik J, de Vries J (2004) Management of aquifer recharge and subsurface storage. A promising option to cope with increasing needs. Netherlands National Committee for the IAH in cooperation with Netherlands Hydrological Society
34. Gale I, Dillon P (2005) Strategies for managed aquifer recharge (MAR) in semi-arid areas. UNESCO Paris
35. Pervin M (2015) Potential and challenges of managed aquifer recharge in an overexploited aquifer of Dhaka City
36. Wang W, Zhou Y, Sun X, Wang W (2014) Development of managed aquifer recharge in China. Bull Geol Soc Am 125(2):227–233
37. Gammie G, De Bievre B, Guevara O (2015) Assessing green interventions for the water supply of Lima, Peru
38. Zhang H, Xu Y, Kanyerere T (2020) A review of managed aquifer recharge: historical development, current situation and perspectives. Appl Geochem 118:102887

39. Theis CV (1935) The relation between the lowering of the piezometric surface and the rate and duration of discharge of a well using groundwater storage. Trans Am Geophys Union 16(2):519–524
40. van Lidth de Jeude E (2016) Managed aquifer recharge: opportunities and barriers
41. Weeks E (2002) A historical overview of hydrologic studies of artificial recharge in the US geological survey. In: US geological survey artificial recharge workshop proceedings
42. Huisman L, Olsthoorn T (1983) Artificial groundwater recharge. Pitman London
43. David R, Pyne G, Hutchinson A (2005) Aquifer storage recovery: a guide to groundwater recharge through wells 28(2):172–173
44. Charlesworth P, Narayan K, Bristow KL, Lowis B, Laidlow G, McGowan R (2020) The Burdekin Delta—Australia's oldest artificial recharge scheme. In: Management of aquifer recharge for sustainability. CRC Press, pp 347–352
45. Sherif M, Sefelnasr A, Al Rashed M, Alshamsi D, Zaidi FK, Alghafli K, Baig F, Al-Turbak A, Alfaifi H, Loni OA, Ahamed MB (2023) A review of managed aquifer recharge potential in the Middle East and North Africa Region with examples from the Kingdom of Saudi Arabia and the United Arab Emirates. Water 15(4):742. https://doi.org/10.3390/w15040742
46. Escalante EF, Gil RC, San Miguel Fraile MÁ, Serrano FS (2014) Economic assessment of opportunities for managed aquifer recharge techniques in Spain using an advanced geographic information system (GIS) 6(7):2021
47. Gruetzmacher G, Kumar PS (2012) Introduction to managed aquifer recharge (MAR)—Overview of schemes and settings worldwide. In: Paper presented at the conference on managed aquifer recharge: methods, hydrogeological requirements, post and pre-treatment systems. https://www.researchgate.net/publication/234657314
48. Sprenger C, Hartog N, Hernández M, Vilanova E, Grützmacher G, Scheibler F, Hannappel S (2017) Inventory of managed aquifer recharge sites in Europe: historical development, current situation and perspectives. Hydrogeol J 25(6):1909–1922. https://doi.org/10.1007/s10040-017-1554-8
49. Maliva R, Missimer T (2012) Managed aquifer recharge. In: Arid lands water evaluation and management. Springer, pp 559–630
50. Standen K, Costa LR, Monteiro JPJW (2020) In-channel managed aquifer recharge: a review of current development worldwide and future potential in Europe 12(11):3099
51. Bai J, Su X, Wang J, Lyu H, Gao R, Lu SJ (2020) Multi-isotope constraints on biogeochemical processes during bank filtration: a case study of the Liao River, Northeast China 122:104762
52. Bertrand G, de Paiva A, de Araújo Freitas J, da Silva Pereira Cabral J, Veras Albuquerque T, de Carvalho Filho JJ (2021) River bank filtration in tropical metropoles: integrated evaluation of physical, geochemical and biochemical interactions in Recife, NE Brazil 1–16
53. Chen Y, Su X, Wan Y, Lyu H, Dong W, Shi Y, Zhang YJ (2022) Nitrogen biogeochemical reactions during bank filtration constrained by hydrogeochemical and Songhua River, NE China 140:105272
54. Schijven JF, Hassanizadeh SM (2000) Removal of viruses by soil passage: overview of modeling, processes, and parameters. Crit Rev Environ Sci Technol 30(1):49–127
55. Sprenger C, Lorenzen G, Hülshoff I, Grützmacher G, Ronghang M, Pekdeger A (2011) Vulnerability of bank filtration systems to climate change. Sci Total Environ 409(4):655–663
56. Pang L (2009) Microbial removal rates in subsurface media estimated from published studies of field experiments and large intact soil cores. J Environ Qual 38(4):1531–1559
57. Jiang X, Zhou J, Zhu M, He W, Yu G (2001) Charge characteristics on the clay surface with interacting electric double layers. Soil Sci 166(4):249–254
58. Nwankwo CB, Hoque MA, Islam MA, Dewan AJES (2020) Groundwater constituents and trace elements in the basement aquifers of Africa and sedimentary aquifers of Asia: medical hydrogeology of drinking water minerals and toxicants 4(2):369–384
59. Rozemeijer J, Noordhuis R, Ouwerkerk K, Pires MD, Blauw A, Hooijboer A, van Oldenborgh GJJSOTTE (2021) Climate variability effects on eutrophication of groundwater, lakes, rivers, and coastal waters in the Netherlands 771:145366

60. Derx J, Blaschke A, Blöschl GJAiWR (2010) Three-dimensional flow patterns at the river-aquifer interface—A case study at the Danube 33(11):1375–1387
61. Derx J, Blaschke A, Farnleitner A, Pang L, Blöschl G, Schijven J (2013) Effects of fluctuations in river water level on virus removal by bank filtration and aquifer passage—A scenario analysis. J Contam Hydrol 147:34–44
62. Chakraborty SK (2001) Land-use changes: floodplains, dams, and reservoirs–integrated river basins management. In: Riverine ecology, vol 2. Springer, pp 531–607
63. Sonthiphand P, Ruangroengkulrith S, Mhuantong W, Charoensawan V, Chotpantarat S, Boonkaewwan S (2019) Metagenomic insights into microbial diversity in a groundwater basin impacted by a variety of anthropogenic activities. Environ Sci Poll Res 26(26):26765–26781
64. Chambers L, Gooddy D, Binley A JGF (2019) Use and application of CFC-11, CFC-12, CFC-113 and SF6 as environmental tracers of groundwater residence time: a review 10(5):1643–1652
65. Akter A (2022) Groundwater recharge. In: Rainwater harvesting—Building a water smart city. Springer, pp 191–214
66. Kim R, Lee S, Lee J, Kim Y (2007) Design of rainwater management system for eco-housing complex. In: Paper presented at the Proceedings of the rainwater and urban design conference
67. Maryono A, Prasetya A, Sembada P, Basyarudin A (2022) The effect of rainwater injection into dug wells on the changing of groundwater level in Kusumanegara Region of Yogyakarta City. In: Paper presented at the IOP conference series: earth and environmental science
68. Zhang D, Ding X, Liu J, Mei C (2022) Review on mechanism and technical measures of urban rainwater harvesting. In: Paper presented at the IOP conference series: earth and environmental science
69. Bahrami N, Nikoo MR, Al-Rawas G, Al-Wardy M, Gandomi AHJJH (2022) Reservoir optimal operation with an integrated approach for managing floods and droughts using NSGA-III and prospect behavioral theory 127961
70. Valipour M (2016) Optimization of neural networks for precipitation analysis in a humid region to detect drought and wet year alarms. Meteorol Appl 23(1):91–100
71. Hiscock KM, Grischek T (2002) Attenuation of groundwater pollution by bank filtration. J Hydrol 266(3):139–144
72. Besser H, Dhaouadi L, Hamed YJEMJfEI (2022) Groundwater quality evolution in the agro-based areas of southern Tunisia: environmental risks of emerging farming practices 7(1):65–78
73. El Osta M, Masoud M, Alqarawy A, Elsayed S, Gad MJW (2022) Groundwater suitability for drinking and irrigation using water quality indices and multivariate modeling in Makkah Al-Mukarramah Province, Saudi Arabia 14(3):483
74. Bachtouli S, Comte J-CJJCR (2019) Regional-scale analysis of the effect of managed aquifer recharge on saltwater intrusion in irrigated coastal aquifers: long-term groundwater observations and model simulations in NE Tunisia 35(1):91–109
75. Daher W, Pistre S, Kneppers A, Bakalowicz M, Najem WJJH (2011) Karst and artificial recharge: theoretical and practical problems: a preliminary approach to artificial recharge assessment 408(3–4):189–202
76. Van Ginkel M, Olsthoorn T, Smidt E, Darwish R, Rashwan S (2009) Fresh storage salines extraction (FSSE) wells, feasibility of freshwater storage in saline aquifer with a focus on the Red Sea coast, Egypt. In: Paper presented at the Achieving groundwater supply sustainability & reliability through managed aquifer recharge, Proceedings of the 7th international symposium on managed aquifer recharge, ISMAR7, Abu Dhabi, UAE
77. Martinsen G, Bessiere H, Caballero Y, Koch J, Collados-Lara AJ, Mansour M, Sallasmaa O, Pulido-Velazquez D, Williams NH, Zaadnoordijk WJ, Stisen S (2022) Developing a pan-European high-resolution groundwater recharge map–Combining satellite data and national survey data using machine learning. Sci Total Environ 822:153464
78. Ajjur SB, Baalousha HMJWI (2001) A review on implementing managed aquifer recharge in the Middle East and North Africa region: methods, progress and challenges 46(4):578–604
79. Swain S, Taloor AK, Dhal L, Sahoo S, Al-Ansari NJAWS (2022) Impact of climate change on groundwater hydrology: a comprehensive review and current status of the Indian hydrogeology 12(6):1–25

80. Leng G, Huang M, Tang Q, Leung LR (2015) A modeling study of irrigation effects on global surface water and groundwater resources under a changing climate 7(3):1285–1304. https://doi.org/10.1002/2015MS000437
81. Woldearegay K, Tamene L, van Steenbergen F, Mekonnen K (2022) Groundwater recharge through landscape restoration and surface water harvesting for climate resilience: the case of upper Tekeze river basin, Northern Ethiopia. In: Groundwater for sustainable livelihoods and equitable growth. CRC Press, pp 135–157
82. Leng G, Huang M, Tang Q, Gao H, Leung LR (2014) Modeling the effects of groundwater-fed irrigation on terrestrial hydrology over the Conterminous United States. J Hydrometeorol 15(3):957–972. https://doi.org/10.1175/jhm-d-13-049.1
83. Pfeiffer L, Lin CYC (2014) Does efficient irrigation technology lead to reduced groundwater extraction? Empirical evidence. J Environ Econ Manag 67(2):189–208. https://doi.org/10.1016/j.jeem.2013.12.002
84. Khaled Abdella Ahmed A, Shalaby M, Negim O, Abdel-Wahed TJSEJ (2022) Comparative study of the Egyptian code for reusing treated wastewater for agriculture 2(1):1–14
85. Chen S, Wu W, Hu K, Li W (2010) The effects of land use change and irrigation water resource on nitrate contamination in shallow groundwater at county scale. Ecolo Complex 7(2):131–138. https://doi.org/10.1016/j.ecocom.2010.03.003
86. Su Z, Wu J, He X, Elumalai VJE&H (2020) Temporal changes of groundwater quality within the groundwater depression cone and prediction of confined groundwater salinity using Grey Markov model in Yinchuan area of northwest China 12(3):447–468

Conclusions and Recommendations for "Managed Aquifer Recharge in the MENA Countries"

Mustafa El-Rawy and Abdelazim Negm

Abstract The groundwater resources in the MENA region are currently characterized by challenges related to assessment, quality, management, and sustainability. In most MENA countries, groundwater serves as the primary source of fresh water and is utilized for irrigation, domestic, and industrial needs. Adapting to upcoming challenges is key in managing groundwater resources in the MENA region through Managed Aquifer Recharge (MAR). This chapter acknowledges that achieving this goal requires MAR to incorporate integrated flexibility, thorough research, and practical applications. Throughout this book project, editorial teams have pinpointed areas for potential enhancements. Drawing on the authors' conclusions and insights, this chapter presents several recommendations, offering guidance for future researchers to extend the work beyond the scope of this book.

1 Introduction

The Middle East and North Africa (MENA) region, characterized by high temperatures and significant evapotranspiration, relies on various water sources like surface and groundwater, and unconventional methods such as desalination and reused wastewater [1]. Climate change exacerbates these challenges with increasing temperatures and decreasing rainfall, threatening agricultural productivity and intensifying the water scarcity issue. For instance, Tunisia's grain yields could drop by 5–11% by

M. El-Rawy (✉)
Civil Engineering Department, Faculty of Engineering, Minia University, Minia 61111, Egypt
e-mail: mustafa.elrawy@mu.edu.eg

Civil Engineering Department, College of Engineering, Shaqra University, Dawadmi 11911, Saudi Arabia

A. Negm
Water and Water Structures Department Engineering, Faculty of Engineering, Zagazig University, Zagazig 44519, Egypt
e-mail: amnegm@zu.edu.eg

M. El-Rawy and A. Negm (eds.), *Managed Aquifer Recharge in MENA Countries*, Earth and Environmental Sciences Library,
https://doi.org/10.1007/978-3-031-58764-1_11

2050 [1]. The growing population, which is expected to reach 0.5 billion by 2030, further strains the limited water resources, primarily used in agriculture [2].

The current state of water resources in the MENA region is characterized by extreme water stress, which is expected to intensify significantly by 2050 [3]. The region has topped global rankings of water-stressed regions, with every MENA nation predicted to suffer from extreme water scarcity by the mid-century. The demand for water is already outstripping the renewable supply at a worrying rate, with about 83% of the population living under extreme water stress. This situation is exacerbated by the fact that over 80% of water supplies are used for irrigation, livestock, industry, and domestic needs. Even a short drought could have severe consequences, depleting water supplies and forcing governments to cut water to residents. It predicts a significant increase in water demand by 2050, with total water shortage escalating to 199 cubic kilometers per year. This shortage is primarily driven by a 50% increase in demand and a 12% decrease in supply. About 22% of this shortage to climate change and 78% to socio-economic factors [3]. Climate change is anticipated to profoundly affect water resources and reduce agricultural land, potentially intensifying existing global challenges, especially in the MENA region [4–7]. This shift could lead to more severe water scarcity and further constrain agricultural production in an area already facing environmental and socio-economic pressures.

Groundwater plays a pivotal role in fulfilling a substantial portion of global water demands, a fact that is particularly true in many regions of the world. In response to increasing population pressures, numerous developing countries are heavily relying on groundwater to support their ambitious water resource strategies, especially for expanding agricultural activities. Notably, in several countries globally, groundwater accounts for over 70% of total water consumption, highlighting its critical importance in meeting agricultural and domestic water needs [8]. This dependency underscores the need for sustainable management and conservation of groundwater resources to ensure long-term water security. Groundwater is considered the main source of water resources in the majority of the MENA countries [11]. This situation underscores the urgent need for effective water management and innovative solutions like aquifer storage in the MENA region to address the imbalance between water demand and supply.

Managed Aquifer Recharge (MAR) is becoming an increasingly important strategy for addressing water scarcity in the MENA region. This region, characterized by its arid climate and growing water demands due to population growth and agricultural needs, faces significant challenges in water resource management [12]. MAR is a process where water is intentionally injected or infiltrated into aquifers to replenish groundwater supplies. This method is particularly useful in the MENA region for several reasons [13]. Firstly, it helps in mitigating the over-extraction of groundwater, a common issue in the region. Secondly, it can store excess water during times of abundance (like during rare heavy rainfalls) for use during droughts. Thirdly, MAR can enhance water quality through natural filtration processes as water percolates through soil and rock layers [14].

Numerous research contributions have explored the potential for implementing Managed Aquifer Recharge (MAR) in countries such as Saudi Arabia, the United

Arab Emirates [13], the Sultanate of Oman [15–17], Jordan [18], and Egypt [14, 19], with a focus on enhancing water security. These studies have been pivotal in showcasing the escalating interest and practical deployment of MAR practices in these regions.

The book delves into a variety of MAR methods and their feasibility for application in the MENA region, an area particularly challenged by water scarcity and aridity. MAR methods are critical for enhancing groundwater resources and essential for sustainable water management in such arid environments. These methods are explored in detail including infiltration basins, injection wells, constructed wetlands, recharge ponds, and riverbank filtration. Each technique has unique advantages and suitability depending on local conditions like geology, soil types, and water availability.

Infiltration basins are effective in areas with permeable soils, allowing surface water to naturally percolate into aquifers. Injection wells, on the other hand, are suitable for regions with less permeable soils, directly introducing water into the aquifer. Constructed wetlands and recharge ponds serve dual purposes, both recharging groundwater and providing natural water treatment. Riverbank filtration is a viable option near rivers, leveraging natural filtration processes.

The feasibility of these MAR methods in the MENA region is contingent upon several factors, including the availability of water for recharge, local hydrogeological conditions, and socio-economic aspects like land usage and cost. The book likely provides case studies or examples from the MENA region, illustrating the practical application and outcomes of these MAR methods, thus offering valuable insights into sustainable water management strategies tailored to the unique challenges of this arid region.

The following section provides the book chapters' main conclusions and recommendations for researchers and decision-makers. This chapter's findings and recommendations are built on the data/information provided in the book.

2 Conclusions

The conclusion of the book provides a comprehensive analysis of MAR technologies in the MENA region, focusing on the critical need for sustainable and innovative approaches in light of the region's unique challenges.

The summary extends as follows:

The MENA region faces significant water resource challenges due to its arid climate and limited freshwater availability. With less than 1% of the world's freshwater and about 6% of its population, it is one of the most water-scarce regions globally. Low average annual precipitation, often below 200 mm, exacerbates this scarcity. The overuse of groundwater, a primary water source, is alarming, notably in countries like Saudi Arabia. The region's rapidly growing population, projected to increase from its current 562.71 million, along with climate change predictions of decreased rainfall, add further pressure on water resources. Agriculture, consuming up to 90% of the water in the MENA region, further strains these limited resources. In response to these challenges, the region has turned to solutions like desalination, notably in the UAE, and desalination and wastewater reuse, as seen in Egypt. However, these methods are energy-intensive and have environmental implications. As a sustainable alternative, MAR is gaining traction. MAR techniques, including soil aquifer recharge, rainwater harvesting, and riverbank filtration, offer cost-effective and environmentally sustainable means to replenish and preserve groundwater, aiding in the sustainable management of the region's water reserves.

The book underscores the critical importance of groundwater in the NENA region, which faces significant challenges due to overexploitation and pollution. These issues affect the water quantity and quality and harm local ecosystems and economies. To address this, the book advocates for efficient governance and the adoption of innovative technologies. Isotopic techniques are highlighted for their unique ability to identify and quantify groundwater recharge, especially in arid and semi-arid areas. The use of specific isotopes like Oxygen-18, Carbon-14, Bromine, and Krypton-81 has proven effective in various applications across the region, from measuring groundwater recharge and age to detecting seawater intrusion, thereby aiding in the informed management and protection of these vital water resources. Additionally, the isotope of krypton (Kr-81) has been employed in the management of cross-border aquifer systems, particularly within the Nubian Sandstone Aquifer. This application estimates the quantity of groundwater extracted from wells located near oases in Egypt's western desert.

Rainwater Harvesting (RWH) and Managed Aquifer Recharge (MAR): RWH is presented as a dual-purpose solution for enhancing water security and mitigating flood risks. This technique is particularly instrumental in arid or semi-arid regions, where it is used to collect, store, and conserve local surface runoff, mainly for agricultural purposes. RWH is portrayed as a cost-effective and scalable alternative to large infrastructure projects, providing a reliable source of freshwater and reducing the strain on groundwater resources. The book comprehensively discusses various RWH techniques, focusing on their adaptability in different environments and their capacity to mitigate flood risks. Despite facing obstacles such as limited hydrological data, rapid urbanization, and variable climate conditions, RWH is recognized for its significant potential in MAR and in advancing water sustainability, particularly in the MENA region. It also plays a vital role in enhancing agricultural productivity worldwide.

Soil Aquifer Treatment (SAT): SAT is gaining attention as an effective means for wastewater reclamation. The method involves using partially treated wastewater for groundwater recharge, thus playing a key role in water conservation. The effectiveness of SAT depends on several factors, including site characteristics, initial wastewater quality, and system design. The book emphasizes the need for continuous monitoring and adjustment to ensure optimal water quality and treatment efficacy. SAT is particularly beneficial in arid regions like the MENA region, providing an alternative water source for irrigation and other uses.

River Bank Infiltration (RBF): RBF emerges as a promising, cost-effective method for obtaining high-quality water. The technique involves drawing water from rivers into adjacent aquifers, effectively filtering out contaminants. The book highlights the importance of careful site selection and system design to manage the challenges, particularly the high concentration of metals in pumped water. The book demonstrates the effective implementation of RBF in alluvial aquifers adjacent to the Nile River in Egypt, the Zarqa River in Jordan, and the Sirwan Al-Kufa Rivers in Iraq, underscoring the effectiveness of this method in the MENA region. Overall, RBF presents a promising, low-cost approach to alleviating water shortages.

Examples of Successful Managed Aquifer Recharge (MAR) in the MENA Region

Research and data underline the high efficiency of RBF in producing drinking water from the Nile River. RBF can operate as an independent process or as an alternative to direct Nile water extraction, complemented by conventional treatment methods. A key to its success is the continuous operation of RBF wells, which ensures stable water quality by reducing fluctuations and limiting the intake of ambient groundwater that may contain higher levels of iron and manganese. The most effective sites for RBF are identified along areas of full Nile River penetration, such as islands and river bends, where natural geological features favor filtration. RBF stands out as a significant, dependable water source for Upper Egypt, notable for its low operational costs and high-quality output. Overall, the book emphasizes that RBF, through the natural filtration capabilities of riverbanks, could be instrumental in improving the drinking water supply in Egypt, particularly in Upper Egypt, provided there is careful site selection aligned with local hydrogeological conditions.

In Oman, where water scarcity is a critical challenge due to its arid climate, managed aquifer recharge (MAR) has shown promising results in sustainable water management. The country, primarily reliant on irregular rainfall and wadis for natural water sources, has embraced MAR and aquifer storage and recovery (ASR), using treated wastewater to augment its water supply. This strategy has been particularly impactful in coastal areas, addressing salinity issues and enhancing water quality. A notable success is the Salalah Central Sewage Treatment Plant's MAR project, which effectively reduced seawater intrusion from 3.4 to 2.7 km and increased the groundwater level near the injection wells. These results highlight MAR's effectiveness in improving water security and resource management in Oman's challenging terrain.

Groundwater Management in Egypt and Pakistan: The book explores the groundwater management challenges faced by Egypt and Pakistan. Both countries heavily depend on large irrigation systems (the Nile in Egypt and the Indus Basin in Pakistan) that interact closely with groundwater resources. The book stresses the need for artificial recharge and effective management to sustain groundwater availability.

Environmental and Groundwater Quality Risks of Aquifer Recharge with Treated Wastewater

The book also provides a comprehensive overview of the potential environmental, health, societal, and economic risks associated with aquifer recharge, such as groundwater quality degradation and accidental contamination leading to socio-political issues. Despite these risks, the book points out that these risks can be effectively mitigated with effective management and adherence to established quality standards, such as those set by the WHO. This makes aquifer recharge a feasible option for enhancing water resources. Emphasis is placed on the importance of managing MAR systems effectively and conducting ongoing monitoring to reduce health risks. Additionally, the book notes the advantages of utilizing the plant nutrients present in wastewater and the importance of providing additional treatment to the wastewater before it is recharged into the groundwater.

The book categorizes various aquifer recharge techniques and examines their impact on groundwater quality. The book discusses the challenges posed by climate change, urbanization, and population growth, and categorizes AR methods into well or shaft and borehole recharge, runoff harvesting, spreading techniques, and bank filtration. The effectiveness of these methods is influenced by factors such as soil properties, water quality, and climatic conditions. Emphasizing the need for adaptive policies and adherence to water quality standards, the chapter particularly focuses on the safe use of Treated Wastewater (TWW) in MAR, provided specific quality measures are met. Overall, the chapter highlights the significance of AR in sustainable water management and the protection of groundwater quality for future generations.

In summary, the book highlights the critical need for innovative and sustainable water management practices in the NENA region, emphasizing techniques like RWH, SAT, RBF, and MAR as key strategies for ensuring long-term water availability and quality in the face of arid climates, urbanization, and increasing water demand.

3 Recommendations

Adapting to upcoming challenges is key in managing groundwater resources in the MENA Region through MAR. This chapter acknowledges that achieving this goal requires MAR to incorporate integrated flexibility, thorough research, and practical applications. Throughout this book project, editorial teams have pinpointed areas for potential enhancements. Drawing on the authors' conclusions and insights, this chapter presents several recommendations, offering guidance for future researchers to extend the work beyond the scope of this book.

The following recommendations aim to provide a comprehensive roadmap for enhancing the effectiveness and sustainability of MAR projects in the MENA region.

- Implement policies for sustainable water management emphasizing efficient use and controlled groundwater extraction, and develop allocation strategies to balance water demand with supply.
- Upgrade and invest in water infrastructure to minimize leakages, improve distribution, and increase MAR facilities, such as dams, reservoirs, and rainwater harvesting systems.
- Promote water-saving technologies in agriculture, industry, and households, including drip irrigation and efficient water fixtures, to reduce wastage.

- Develop advanced nuclear tracer techniques, like isotopes tritium or carbon-14, for precise groundwater monitoring in the MENA region.
- Foster collaborative research and integrate these techniques with technologies like GIS and AI for comprehensive groundwater management.
- Implement environmental impact assessments to ensure sustainable use of nuclear tracers in water management.
- Conduct pilot projects and case studies in the MENA region to demonstrate rainwater harvesting system effectiveness and identify best practices.
- Invest in technology and research to enhance the efficiency and cost-effectiveness of rainwater harvesting systems, with a focus on innovative designs for maximizing collection and minimizing losses.
- Establish robust monitoring and evaluation mechanisms for assessing the impact of rainwater harvesting on groundwater, soil moisture, and flood mitigation.
- Adapt rainwater harvesting systems to address climate change impacts, especially water scarcity and increased flood risks.
- Build partnerships with governments, NGOs, international agencies, and the private sector to support the implementation and growth of rainwater harvesting initiatives.
- Prioritize sustainability and minimal environmental impact in rainwater harvesting practices.
- Enhance data collection and analysis to understand rainfall patterns and water table levels, optimizing rainwater harvesting effectiveness in the MENA region.
- Integrate advanced treatment technologies in SAT (Soil Aquifer Treatment) processes to enhance contaminant removal from treated wastewater, ensuring better water quality for aquifer recharge, while also designing these systems for sustainability, focusing on energy efficiency and long-term viability.
- Implement comprehensive monitoring programs for regular water quality assessment in aquifers, ensuring compliance with health and environmental standards, and developing stringent safety protocols to mitigate environmental and health risks associated with using treated wastewater.
- Conduct ongoing research to deepen understanding of the interactions between treated wastewater, soil, and aquifers in the MENA region, thereby optimizing SAT processes for various geological and climatic conditions and designing these systems to be adaptable to climate change impacts.
- Foster collaborative partnerships between public and private sectors to pool resources, expertise, and innovation for effective SAT project implementation and integration of these systems into broader water resource management strategies, enhancing overall water security and sustainability in the MENA region.
- Perform comprehensive hydrogeological studies to select ideal locations for River Bank Filtration (RBF) in the MENA region, considering river flow, soil composition, and pollutant levels to ensure effectiveness.
- Utilize advanced monitoring and filtration technologies in RBF systems for optimal contaminant removal, targeting specific regional water quality issues like salinity and heavy metals, and design these systems for sustainability with energy efficiency and low maintenance needs.

- Enhance Oman's water management by upgrading desalination plants for drinking water and improving wastewater treatment facilities, coupled with modernizing irrigation through advanced rainwater harvesting. Implement mandatory sustainable water use laws across all sectors and focus on integrating Managed Aquifer Recharge and Aquifer Storage and Recovery techniques for groundwater quality.
- Expand River Filtration Technology in Upper Egypt along the Nile, enhancing water resources, and implementing continuous, comprehensive monitoring to ensure water quality and safety.
- To mitigate the environmental risks associated with Aquifer Recharge using partially treated wastewater, it is recommended to encourage developing societies to use cost-effective tracers like dyes and chloride in MAR projects to monitor the status and movement of recharged water, while developed countries can opt for more advanced tracers like radon and stable isotopes of water. The choice of tracers should align with the region's economic capabilities and the specific requirements of each MAR project.

Finally, for effective implementation of MAR in the MENA region, it is important to consider factors like the identification of suitable recharge sites, the quality of water to be recharged, and the social, economic, and environmental impacts. Additionally, legal and institutional frameworks need to be in place to manage these initiatives effectively. Collaboration among countries, especially for transboundary aquifers, is also crucial to the success of MAR programs.

References

1. Haddaoui I, Mateo-Sagasta J (2021) A review on occurrence of emerging pollutants in waters of the MENA region. Environ Sci Pollut Res 28:68090–68110. https://doi.org/10.1007/s11356-021-16558-8
2. The World Bank (2023) The World Bank database. https://data.worldbank.org/indicator/SP.POP.TOTL?locations=ZQ. Accessed 10 Sept 2023
3. Droogers P, Immerzeel WW, Terink W, Hoogeveen J, Bierkens MF, Van Beek LP, Debele B (2012) Water resources trends in Middle East and North Africa towards 2050. Hydrol Earth Syst Sci 16:3101–3114. https://doi.org/10.5194/hess-16-3101-2012
4. El-Rawy M, Batelaan O, Al-Arifi N, Alotaibi A, Abdalla F, Gabr ME (2023) Climate change impacts on water resources in arid and semi-arid regions: a case study in Saudi Arabia. Water 15(3):606. https://doi.org/10.3390/w15030606
5. El-Rawy M, Fathi H, Zijl W, Alshehri F, Almadani S, Zaidi FK, Aldawsri M, Gabr ME (2023) Potential effects of climate change on agricultural water resources in Riyadh Region, Saudi Arabia. Sustainability 15(12):9513. https://doi.org/10.3390/su15129513
6. Omran ES, Negm AM (2020) Climate change impacts on agriculture and food security in Egypt. Springer, Cham. https://doi.org/10.1007/978-3-030-41629-4
7. Al-Maktoumi A, Zekri S, El-Rawy M, Abdalla O, Al-Wardy M, Al-Rawas G, Charabi Y (2018) Assessment of the impact of climate change on coastal aquifers in Oman. Arab J Geosci 11:501. https://doi.org/10.1007/s12517-018-3858-y
8. Nriagu JO (2019) Encyclopedia of environmental health. 2nd edn. Elsevier
9. El-Rawy M, Abdalla F, El Alfy M, 2020. Water resources in Egypt. In: The geology of Egypt. Springer, Cham, pp 687–711. https://doi.org/10.1007/978-3-030-15265-9_18

10. El-Rawy M, Makhloof AA, Hashem MD, Eltarabily MG (2021) Groundwater management of quaternary aquifer of the Nile Valley under different recharge and discharge scenarios: a case study Assiut governorate, Egypt. Ain Shams Eng J 12(3):2563–2574. https://doi.org/10.1016/j.asej.2021.02.023
11. Negm AM, El-Rawy M (eds) (2022) Sustainability of groundwater in the Nile Valley, Egypt. Springer Nature, Cham. https://doi.org/10.1007/978-3-031-12676-5
12. Zekri S, Al-Maamari A (2020) An overview of the water sector in MENA Region. In: Zekri S (eds) Water policies in MENA Countries. Global Issues in Water Policy, vol 23. Springer, Cham. https://doi.org/10.1007/978-3-030-29274-4_1
13. Sherif M, Sefelnasr A, Al Rashed M, Alshamsi D, Zaidi FK, Alghafli K, Baig F, Al-Turbak A, Alfaifi H, Loni OA et al (2023) A review of managed aquifer recharge potential in the Middle East and North Africa Region with examples from the Kingdom of Saudi Arabia and the United Arab Emirates. Water 15(4):742. https://doi.org/10.3390/w15040742
14. Amin HM, Gad AA, El-Rawy M (2021) Abdelghany UA, Sadeek RA. Assessment of wastewater contaminant concentration through the vadose zone in a soil aquifer treatment system. Appl Ecol Environ Res 19(3):2385–2403. https://doi.org/10.15666/aeer/1903_23852403
15. Al-Maktoumi A, El-Rawy M, Zekri S (2016) Management options for a multipurpose coastal aquifer in Oman. Arab J Geosci 9:636. https://doi.org/10.1007/s12517-016-2661-x
16. El-Rawy M, Al-Maktoumi A, Zekri S, Abdalla O, Al-Abri R (2019) Hydrological and economic feasibility of mitigating a stressed coastal aquifer using managed aquifer recharge: a case study of Jamma aquifer, Oman. J Arid Land 11:148–159. https://doi.org/10.1007/s40333-019-0093-7
17. Al-Maktoumi A, Zekri S, El-Rawy M, Abdalla O, Al-Abri R, Triki C, Bazargan-Lari MR (2020) Aquifer storage and recovery, and managed aquifer recharge of reclaimed water for management of coastal aquifers. Desalin Water Treat 176:67–77
18. El-Rawy M, Zlotnik VA, Al-Raggad M, Al-Maktoumi A, Kacimov A, Abdalla O (2016) Conjunctive use of groundwater and surface water resources with aquifer recharge by treated wastewater: evaluation of management scenarios in the Zarqa River Basin, Jordan. Environ Earth Sci 75:1–21. https://doi.org/10.1007/s12665-016-5946-1
19. Abdelrady A, Sharma S, Sefelnasr A, El-Rawy M, Kennedy M (2020) Analysis of the performance of bank filtration for water supply in arid climates: case study in Egypt. Water 12(6):1816. https://doi.org/10.3390/w12061816

SPRINGER NATURE

GPSR Compliance

The European Union's (EU) General Product Safety Regulation (GPSR) is a set of rules that requires consumer products to be safe and our obligations to ensure this.

If you have any concerns about our products, you can contact us on ProductSafety@springernature.com

In case Publisher is established outside the EU, the EU authorized representative is:

Springer Nature Customer Service Center GmbH
Europaplatz 3
69115 Heidelberg, Germany

The manufacturer's authorised representative in the EU is Springer Nature Customer Service Centre GmbH, Europaplatz 3, 69115 Heidelberg, Germany. If you have any concerns regarding our products, please contact ProductSafety@springernature.com

Printed and bound by CPI Group (UK) Ltd, Croydon, CR0 4YY
23/09/2026
02221546-0003